FAILED
TECHNOLOGY

TRUE STORIES OF TECHNOLOGICAL DISASTERS

FAILED TECHNOLOGY

TRUE STORIES OF TECHNOLOGICAL DISASTERS

Volume 1

Fran Locher Freiman
& Neil Schlager

An imprint of Gale Research Inc.,
an International Thomson Publishing Company

I(T)P

NEW YORK • LONDON • BONN • BOSTON • DETROIT • MADRID
MELBOURNE • MEXICO CITY • PARIS • SINGAPORE • TOKYO
TORONTO • WASHINGTON • ALBANY NY • BELMONT CA • CINCINNATI OH

FAILED
TECHNOLOGY
TRUE STORIES OF TECHNOLOGICAL DISASTERS

Fran Locher Freiman and Neil Schlager, *Editors*

Staff

Carol DeKane Nagel, *U·X·L Developmental Editor*
Thomas L. Romig, *U·X·L Publisher*

Mary Kelley, *Production Associate*
Evi Seoud, *Assistant Production Manager*
Mary Beth Trimper, *Production Director*

Mark Howell, *Page and Cover Designer*
Cynthia Baldwin, *Art Director*

H. Diane Cooper, *Permissions Associate (Pictures)*
Margaret A. Chamberlain, *Permissions Supervisor (Pictures)*

The Graphix Group, *Typesetter*

Library of Congress Cataloging-in-Publication Data

Freiman, Fran Locher
    Failed technology: true stories of technological disasters / Fran Locher Freiman, Neil
Schlager.
        2 v. (xviii, 392 p.): ill.; 24 cm.
    Includes bibliographical references (v. 1, p. 173-175, v. 2, p. 383-385) and indexes.
    ISBN 0-8103-9794-3 (set). -- ISBN 0-8103-9795-1 (v. 1). -- ISBN 0-8103-9796-X (v. 2).
    1. System failures (Engineering)--Case studies. 2. Disasters--Case studies. I. Freiman, Fran Locher II.
Schlager, Neil, 1966- III. Title.
TA169.5 .F74 1995   96-129100
620.20--dc20   CIP

♾™ This book is printed on acid-free paper that meets the minimum requirements of American National Standard for Information Sciences—Permanence Paper for Printed Library Materials, ANSI Z39.48-1984.

Printed in the United States of America
10 9 8 7 6 5 4 3 2

Contents

Preface

People today can get very enthusiastic about using technology, especially because it mostly works very well and so greatly enhances the quality of our lives. Technological tools are around us everywhere—they wake us up, entertain us, keep our food fresh, help us prepare and cook our meals and wash the dishes afterwards. They take us where we want to go and keep us comfortable while we are on our way. We look forward to the new and improved technologies, and we hope they come in our favorite color and in the latest style.

Our expectations about the technological devices in our lives are so high because we rely on them so completely. We hardly question how all these tools came to exist, and this reliance itself testifies to how well technology works most of the time. But we have all experienced occasions of failed technology, such as when the electrical power goes out or our cable television company has technical problems. We hate being inconvenienced.

But what about when the failure is a major one? Henry Petroski writes in his new book, *Design Paradigms*, "Some of history's most embarrassing moments have come in the dramatic failures of some of the largest machines, structures, and systems ever attempted." It surely was embarrassing to the owners of the steamship *Titanic* when the liner sank, despite the owners' boast that their ship was unsinkable. And NASA was certainly embarrassed when the space shuttle *Challenger* exploded and when the shuttle inquiry revealed the embarrassing facts about the O-ring deficiencies.

However, failed technologies produce far more than embarrassment. They endanger human lives or end them. They disrupt our economic livelihoods. They shock us and set us back in our confidence. They make us question how we came to rely so much on manufactured products. Failures frighten us.

But aside from what we lose in disasters, we also learn from them. All technological development represents the thoughtful application of design expressed in materials. When successful designs are modified, the new design seems improved. However, Petroski reminds us: "Any design change . . . can introduce new failure modes or bring into play latent failure modes. . . . While a structure designed 'the old way' may be perfectly safe, an 'improved' or enlarged design could hold very unpleasant surprises."

Format and Inclusion Criteria

Failed Technology: True Stories of Technological Disasters focuses on 44 unpleasant surprises of the twentieth century. The failed technologies span several fields:

- Ships and Submarines
- Airships, Aircraft, and Spacecraft
- Automobiles
- Dams and Bridges
- Buildings and Other Structures
- Nuclear Plants
- Chemical and Environmental Disasters
- Medical Disasters.

The catastrophes highlighted in *Failed Technology* resulted from poor design, planning, testing, or construction—not from purely natural disasters such as earthquakes and hurricanes nor deliberate human actions such as terrorist bombings. Other disasters are included because of the media coverage they received and because of their impact on public opinion. Many of the failures are well known. But even when they are not dramatic or close to home, they reveal how technology has gone wrong.

The entries in *Failed Technology* have a common format. Each contains major sections in the following pattern:

- Background
- Details of the Disaster
- Impact
- Where to Learn More

Additional headlines highlight the specific incidents that led up to the catastrophe and occurred in its aftermath. The volumes also feature more than 70 photos and illustrations, a chronology of technological disasters, a bibliography, and an cumulative subject index.

Comments and Suggestions

We welcome your comments on this work as well as your suggestions for disasters to be featured in future editions of *Failed Technology: True Stories of Technological Disasters*. Please write: Editors, *Failed Technology*, U·X·L, 835 Penobscot Bldg., Detroit, Michigan 48226-4094; call toll-free: 1-800-877-4253; or fax: 313-961-6348.

Picture Credits

The photographs and illustrations appearing in *Failed Technology: True Stories of Technological Disasters* were received from the following sources:

The Bettmann Archive: pages 4, 6, 74, 88; **AP/Wide World Photos:** pages 9, 25, 29, 39, 54, 118, 129, 134, 136, 150, 155, 162, 186, 193, 198, 227, 237, 251, 261, 266, 276, 302, 315, 319, 321, 361; **UPI/Bettmann Newsphotos:** pages 13, 64, 284, 286, 342; **UPI/Bettmann:** pages 20, 61, 71, 79, 96, 103, 110, 206, 209, 214, 222, 225, 248, 255, 257, 334, 335, 345, 377; **Reuters/Bettmann:** pages 23, 91, 126, 300; **U.S. Navy:** pages 36, 44; **Archive Photos/Lambert:** page 69; **Reprinted by permission of W. W. Norton & Company, Inc.:** page 81; **Copyright 1979 Time Inc., reprinted by permission:** p. 105; **Sovfoto:** page 141; **NASA:** pages 147, 167, 169; **Kyodo News Service:** page 294; **Archive Photos/Orville Logan Snider:** page 312; **Black Star:** page 327; **Peter A. Simon/Phototake, NYC:** page 370.

Chronology of Technological Failures

September 30, 1911	Austin Dam fails
April 15, 1912	R.M.S. *Titanic* sinks
September 3, 1925	U.S.S. *Shenandoah* crashes
October 5, 1930	R-101 crashes
May 6, 1937	*Hindenburg* explodes
May 23, 1939	U.S.S. *Squalus* sinks
1939—	DDT insecticide contamination
November 7, 1940	Tacoma Narrows Bridge collapses
1940-79	Diethylstilbestrol (DES)
1942-80	Love Canal toxic waste site
January 10, 1954	BOAC Comet explodes
April 8, 1954	BOAC Comet explodes
1955—	Minamata Bay mercury poisoning
July 26, 1956	*Andrea Doria* sinks
September 29, 1959	Lockheed Electra crashes
1959-63	Chevrolet Corvairs roll over
1950s-60s	Thalidomide
March 17, 1960	Lockheed Electra crashes
December 16, 1960	United Airlines DC-8 and TWA Constellation collide
January 3, 1961	SL-1 reactor explodes
1961-71	Agent Orange contamination
April 10, 1963	U.S.S. *Thresher* sinks

October 9, 1963	Vaiont Dam landslide
January 27, 1967	Apollo 1 catches fire
April 24, 1967	*Soyuz 1* crashes
May 27, 1968	U.S.S. *Scorpion* lost
1960s—	Silicone-gel implants
April 13, 1970	*Apollo 13* oxygen tank explodes
1971-76	Ford Pintos explode (1971-76 model years)
May 14, 1973	Skylab's meteoroid shield fails
March 28, 1979	Three Mile Island reactor melts down
May 25, 1979	American Airlines DC-10 crashes
1970s	Dalkon Shield Intrauterine Device
November 21, 1980	MGM Grand Hotel fire
March 8, 1981	Tsuruga radioactive waste spills
July 17, 1981	Hyatt Regency Hotel walkways collapse
February 15, 1982	*Ocean Ranger* rig sinks
August 1982	Zilwaukee Bridge fails
December 3, 1984	Union Carbide toxic vapor leak
August 12, 1985	Japan Airlines Boeing 747 crashes
January 28, 1986	*Challenger* explodes
April 26, 1986	Chernobyl reactor explodes
April 5, 1987	Schoharie Creek Bridge collapses
February 24, 1989	United Airlines Boeing 747 explodes
March 24, 1989	Exxon *Valdez* runs aground
October 4, 1992	El Al Boeing 747-200 crashes

Ships and Submarines

R.M.S. *Titanic* sinks

Atlantic Ocean off Newfoundland,
Collided with an iceberg, April 14, 1912
Sank, April 15, 1912

Background

The R.M.S. *Titanic* was on her maiden voyage from Southampton to New York when, at 11:40 P.M. on the night of April 14, 1912, she struck an iceberg off Newfoundland. Just 2 hours and 40 minutes later she sank, and with her perished over 1,500 passengers and crew members. The *Titanic* was the largest and most luxurious ship afloat and was also believed to be "practically unsinkable." First seen to symbolize the triumph of technology, it afterwards became a symbol of human arrogance.

The idea for building the *Titanic* originated in 1907 in the minds of Lord Pirie and J. Bruce Ismay. Pirie headed Harland & Wolff, a shipbuilding firm known to build the sturdiest and best ships in the British Isles. J. Bruce Ismay chaired the White Star Line, which in 1902 had been taken over by a huge trust company, International Mercantile Marine. In effect, the takeover made American financier J. Pierpont Morgan the principal owner of the *Titanic*. Although the venture was a marriage of American money and British technology, the *Titanic* was essentially British: registered as a British ship and manned by British officers, she would revert to the British navy in time of war.

Titanic's builders want the biggest and grandest vessel

The early 1900s was a time of great rivalry among steamship companies. Pirie and Ismay felt that if they couldn't compete with the crack Cunard liners in speed, then they would surpass them in size and grandeur. They also decided to serve the growing immigrant traffic and greatly increased accommodations for steerage, or low-fare, passengers.

A state-of-the-art ocean liner sinks after hitting an iceberg in the frigid North Atlantic.

The R.M.S. *Titanic* as she set out on her voyage from Southampton.

The bow is the forward part of a ship. The stern is the rear end of a ship. Amidships is midway between bow and stern. Aft means near, toward, or in the stern of a ship.

They built the *Titanic* in Belfast alongside her sister ship, the *Olympic*. While the *Titanic* surpassed the *Olympic* in gross tonnage, she was shorter in length. The *Titanic* was launched on May 31, 1911.

The *Titanic* was 882 feet in length, 92 feet in width, and weighed 46,328 gross tons; her 9 steel decks rose as high as an 11-story building. The stability afforded by her vast size was deemed to be one of many safeguards against sinking. She also contained a much greater proportion of steel for her structural parts than previous ships. Also, *Titanic* was built with a double bottom. This was not an innovation, but both skins were heavier and thicker than those previously used in ships. The outer skin of the bottom was a full inch thick.

Double-bottomed steel hull boasted "watertight" compartments

The huge hull was divided by 15 transverse bulkheads, or upright partitions, extending the width of the ship. This created 16 "watertight" compartments, any 2 of which could flood without affecting the safety of the ship. The 6 compartments that contained boilers had their own pumping equipment. The doors between watertight compartments could be closed all at once by a switch on the bridge or individually by crewmen. Through the bulkheads were 5 decks above forward and aft and 4 decks above amidships.

The *Titanic* enjoyed her reputation for being the biggest ship afloat as well as the most elegant. She had the first shipboard swimming pool, a Turkish bath with a gilded cooling room, a gymnasium, a squash court. The dining rooms, staterooms, and common rooms were furnished in various period styles. The first-class cabins were especially opulent, some with coal-burning fireplaces in the sitting rooms and full-size four-poster beds in the bedrooms. The ship had a hospital with a modern operating room, and guests could have their cars when they reached port, for there was a compartment for automobiles and a loading crane for off-loading them.

The port side is the left side of a ship looking forward. The starboard side is the right side of a ship looking forward.

Overconfidence was a sign of the times

There were few detractors of the *Titanic,* probably because the very idea of traveling in such opulence embodied the overconfident spirit of the times. Some observers, though, pointed out that there wasn't a dock in America big enough for the *Titanic* or the *Olympic.* Others felt that the attention to luxury represented an increased expense and no advancement in economic transportation. Still others—ironically, as it turned out—pointed out that a ship so large would concentrate too much wealth and too many lives in a single vessel. They wondered if underwriters would even agree to take on so great a risk.

Details of the Sinking

The evening of April 14, the fifth day of the *Titanic's* maiden voyage, the sea was exceptionally calm and the sky was starry but moonless. The ship cruised at about 21 knots. The crew had received a number of warnings about ice in the region, and Captain E. J. Smith is known to have seen at least four icebergs. He did not alter his speed.

There are several probable reasons he did not reduce the ship's speed:

- Visibility seemed good
- The captain may have been influenced by the *Titanic's* reputation for being "practically unsinkable"

A contemporary drawing of how the *Titanic* may have struck the iceberg.

- Ice fields were a hazard he had much experience with

- He needed to project an aura of "robust captaincy"—a matter of pride among sea captains, especially those with the White Star Line

- The captain was probably trying to make good speed on the maiden voyage

Nevertheless, April is one of the worst months for icebergs. Icebergs are the seaward tips of glaciers that break off, and most come from the west coast of Greenland. About a thousand of these "glacier calves" migrate into the shipping lanes each year.

As the *Titanic* navigated these waters, there were two crew members at their watch. The lookouts were working without binoculars, which were supposed to be standard equipment on the White Star Line. They did not see the iceberg until it was just a quarter mile from the ship. The bow was swung swiftly to port at 11:40 P.M., but it was a futile maneuver. The underwater portion of the iceberg tore through the plating on the starboard bow, aft to perhaps amidships. The collision breached possibly six of the watertight compartments. The bow started to sink. Then more compartments filled with water, which soon sloshed over the tops of the transverse bulkheads.

The collision was scarcely felt

Traveling at 21 knots, the 46,000-ton *Titanic* scarcely felt the impact. The energy absorbed in colliding with the iceberg was insignificant in proportion to the total energy of the vessel. The gravity of the situation was comprehended only gradually, and because of the late hour, many passengers had to be awakened. At 12:20 A.M. came the order to ready the lifeboats.

Great confusion developed during the loading of the lifeboats. The liner was equipped with only 16 lifeboats and 4 emergency rafts—enough for about half of the passengers. The *Titanic*'s lifeboat supply was actually in compliance with British Board of Trade mandates of the time, which calculated the number of lifeboats required according to the tonnage of the ship, not to passenger capacity. The board also permitted reduction in the number of boats for ships deemed to have satisfactory watertight subdivision.

Not enough lifeboats and no lifeboat drills

Having had no lifeboat drills, the crew and passengers did not know which lifeboats they should go to. Furthermore, the officers in charge of loading them were poorly trained. They feared that if they loaded the boats to capacity, the lifeboats would buckle as they were being lowered. They also worried that the davits holding the boats over the side would break. (Ironically neither would have occurred, because the boats and davits had been tested.) As a result the officers sent the boats down only partly loaded, with instructions to come alongside the cargo ports to pick up more passengers. The cargo ports were never opened, however, and many boats went away only partly filled. The boats would have accommodated 1,178 people. Only 711 found place in the boats; 467 other lives might have been saved.

A bulkhead is an
upright partition
separating
compartments
that resist the
pressure of the
seawater. A davit
is a crane
projecting over
the side of a ship.

Titanic sounded a "death rattle" while she sank

As the *Titanic's* bow sank lower and lower, the people left behind on the sinking ship climbed to the stern, from which some jumped into the 28°F water in their life belts. As the survivors in the lifeboats fixed their gaze on the ship, the immense stern reared up almost to the perpendicular and remained still for a few moments—then, at 2:20 A.M. on April 15, 1912—she was swallowed up by the sea. Many survivors reported hearing sounds like thunder, and deep detonations, and a kind of "death rattle" before the stern went down.

Some passengers reported that the great ship broke in two. The explanation generally accepted to explain the thunderous sound was that as the stern rose, the boilers crashed down through the bulkheads. The position of the bow and the stern on the ocean floor—facing in opposite directions about 1,970 feet apart—makes it seem likely that the ship did break in two at or near the surface. As the bow sank and the stern rose, the pressure on the keel probably increased until it snapped.

Searches of the wreckage were made in 1985 by a Franco-American expedition and in 1986 by an American team. The 1986 expedition was led by Dr. Robert D. Ballard of the Woods Hole Oceanographic Institution. The teams located and photographed the wreck in 13,000 feet of water, looking for the iceberg's telltale gash in the hull. Dr. Ballard observed many buckled plates below the waterline—but no gash—which may remain hidden deeper down, where the bow sank with great force into the sediment.

Nearest ship to hear plea was 58 miles away

The state of wireless communication at that time frustrated lifesaving measures. The *Californian* was stopped for the night in ice fields not more than 20 miles away, but by the time the *Titanic's* distress call went out, the *Californian's* wireless operator had closed up for the night—just 15 or 20 minutes earlier. Wireless operators were employed by the Marconi company and did not follow around-the-clock shipboard watches. The *Titanic* succeeded in raising the next closest ship, the *Carpathia*, about 58 miles away. The *Carpathia* picked up the first lifeboat at 4:10 A.M. If there had been lifeboats for everyone aboard the *Titanic*, everyone might have been saved.

Hull's low-grade steel was prone to fracture in icy temperatures

A team of architects and engineers released a report in 1993 arguing that the tragedy was caused—not so much by the collision with the ice-

At 13,000 feet deep the Alvin submarine of the 1986 Woods Hole Oceanographic Institution expedition inspects the *Titanic*'s deck. Researchers from the Institution were unable to locate the gash in the ship's hull caused by the iceberg.

berg—but by the structural weakness of the ship's steel plates. The low-grade steel used on the *Titanic* is subject to brittle fracture—breaking rather than bending in cold temperatures. A better grade of steel might have enabled the ship to withstand the collision or, at the very least, to sink more slowly, which would have increased the time for lifesaving operations. The investigators also suggested that the roar heard by the passengers as the ship sank may have been the steel plates fracturing, not the boilers crashing through the bulkheads.

The brittleness of a material increases as the temperature decreases.

Impact

A U.S. Senate inquiry into the *Titanic* sinking began the day after the *Carpathia* landed in New York with the survivors. William Alden Smith headed the American inquiry. From May through July 1912 a British investigation for the Board of Trade proceeded. It was headed by Lord Mersey as wreck commissioner. An international Safety of Life at Sea (SOLAS) conference met in London from November 1913 to January 1914. Many new rules and regulations resulted from these and later forums.

Lifeboat allocations are revised

New regulations required that lifeboats be supplied according to the number of passengers, not the tonnage of the ship. Lifeboat drills were instituted, to be held soon after a ship set sail. Crew members and passengers would be given specific lifeboat assignments. Seamen assigned to the boats had to have certain training. Standards were set about the kinds of permanent equipment required on-board.

Shipping lanes were moved further south, away from the ice. Ships approaching ice fields were required to slow down or alter their course. (If a ship is traveling at half speed, the reactive blow of a collision is reduced to one fourth, because the energy of a moving mass is proportional to the square of the velocity; e.g., a ship traveling at 20 knots has four times the kinetic energy as that same ship moving at 10 knots.)

Iceberg watch is established

The International Ice Patrol, based in Groton, Connecticut, was established to monitor iceberg movement. Today it employs aircraft with side-looking radar to spot icebergs and also makes computerized predictions about their whereabouts. Since this patrol was formed, no further loss of life has occurred in the North Atlantic shipping lanes due to iceberg collisions.

New wireless regulations require licensing of operators

New regulations were also brought to bear on wireless operations. They stipulated that a ship's wireless operate day and night and have an auxiliary source of power. And because the activity of amateur wireless operators interfered with rescue operations in the period after the *Titanic* sank and before the survivors had been brought to shore, wireless operator licensing became necessary in order to operate a wireless in the United States. This licensing was the beginning of the Federal Communications Commission.

The principal technological impact of the *Titanic* tragedy was how fast it sank. Many believed that it represented state-of-the-art shipbuilding, and people found it hard to believe that it succumbed so quickly. Since the sinking, new regulations about bulkhead subdivision went into effect for the construction of passenger ships, yet no one knows if ships built today—even with high-grade steel construction—could withstand the kind of blow suffered by the *Titanic*.

Losing the *Titanic* and so many of her passengers and crew members produced other, less tangible, effects. After a century of steady technological progress came widespread disillusionment. The sinking of the Titanic was likened to a Greek tragedy in which the hero is destroyed by pride. Ministers preached that God was punishing people for their excessive faith in material progress. There was a profound loss of innocence.

Where to Learn More

Ballard, Robert D., and Rick Archbold. *The Discovery of the "Titanic."* Madison Press Books, 1987.

Davie, Michael. *"Titanic": The Death and Life of a Dream.* New York: Alfred A. Knopf, 1987.

"Scientific Aftermath of the *Titanic* Disaster." *Literary Digest* (May 25, 1912): 1096–1097.

"Tragedy of the *Titanic:* A Complete Story." *New York Times* (April 28, 1912): Sect. 6, 1–8.

Wade, Wyn Craig. *The "Titanic": End of a Dream.* New York: Rawson, Wade, 1979.

"Wreck of the *Titanic*." *Engineering Magazine* (June 1912): 446–449.

Andrea Doria sinks

Off the Massachusetts coast
Collided with the *Stockholm*, July 25, 1956
Sank, July 26, 1956

Fifty-two people lose their lives when radar misreading causes the Italian steamer Andrea Doria *and another steamer, the* Stockholm, *to collide.*

Background

On July 25, 1956, the steamships *Andrea Doria* and *Stockholm* collided near the *Nantucket Lightship* off the Massachusetts coast. Forty-three passengers and crew of the *Andrea Doria* died from the impact of the collision, and 11 hours later the ship sank. The *Doria* was the first big liner to be lost in peacetime since the *Titanic* went down in 1912. On the *Stockholm*, three crew members were never seen again, and several later died from their injuries. The collision was caused primarily by misreading of radar signals. The sinking resulted from faulty ballasting.

The *Andrea Doria* was the flagship of the Società Italia di Navigazione (Italian Line). She was built by Ansaldo of Sestri, near Genoa, Italy, and launched in 1951. *Doria* was the first passenger ship to be built in Italy after World War II, and Genoa was her home port. The ship was 700 feet in length, weighed 29,083 tons, had a service speed of 23 knots, and was known for her graceful lines. On what was to be her last voyage, the *Andrea Doria* left Genoa on July 17, 1956, and was due in New York City on July 26.

The *Stockholm*, owned by the Svenska Amerika Linien (Swedish-American Line), was built in Sweden by Götaverken of Göthenberg and launched in 1943. It was 525 feet overall, weighed 12,644 tons, and had a service speed of 19 knots. On July 25, 1956, the ship was on its first day out, beginning its homeward voyage from New York City to Sweden.

Ships heading toward the *Nantucket Lightship* from opposite directions

Both ships were heading toward the *Nantucket Lightship*, a vessel

The *Andrea Doria* listed heavily before sinking. The ship's severe list (18°, then 22°) rendered the port-side lifeboats useless and made the starboard-side lifeboats hard to access, causing them to swing far out over the water.

anchored 50 miles off Nantucket to guard the shoals off that island. The *Nantucket Lightship* is an important focal point for the sea lanes of the world.

Ships traveling between ports take the shortest distance possible, following a great circle on the earth's surface. A great circle—such as the equator—is an imaginary circle on the Earth established by intersecting a geometric plane through its center. The shortest great circle in the North Atlantic actually tracks north of the *Nantucket Lightship,* but due to weather and navigational hazards, ships cannot always follow their shortest route.

Ships traveling from New York to European ports must pass to the south of the *Nantucket Lightship* before turning north. The major steamship lines had agreed that their eastbound ships would pass in a track 20 miles south of the light vessel, while their westbound ships would pass close to it. However, neither the Italian Line nor the Swedish-American Line was a party to this international agreement.

Details of the Collision

On the afternoon of July 25, 1956, the *Andrea Doria* was eight days out of Genoa and due to dock in New York the next morning. She encountered fog about 150 miles from the *Nantucket Lightship*. Captain Piero Calamai observed the normal precautions for these conditions. He:

- Closed the watertight doors
- Sounded regular prolonged blasts on the siren
- Assigned an officer to keep a constant watch on the radar screen for any sign of approaching ships
- Reduced the ship's speed

Only the last of Captain Calamai's actions could be criticized, for he reduced the *Doria's* speed only 5 percent, from 22 knots to 21 knots. The collision rules call for a "moderate speed," that is, a speed that allows the ship to be stopped within its visibility distance. By 8:00 A.M. the *Doria's* visibility was only half a mile, an impossibly short distance to have to stop her in. However, sea captains bound by schedules customarily made only token reductions in speed in bad weather, while keeping a close watch on their radar.

Blip on radar screen reveals steamers are on collision course

At 10:20 A.M. the westbound *Doria* passed one mile to the south of the light vessel. At about 10:45 second Officer Franchini saw a "blip" on the radar screen. The blip represented a ship that appeared to be not quite due ahead, but fine, or slightly, to starboard.

Captain Calamai checked the radar screen himself and judged that the rapidly approaching ship was "to starboard"—that is, to the right—of his heading marker. The captain felt quite certain that the two ships were on a starboard-to-starboard path, but he thought that the distance between would be uncomfortably small. He gave the order to alter course 4° to port (left).

In so doing, he was going against Rule 18 of the collision rules, which states that "when two power-driven vessels are meeting end on, or nearly end on, so as to involve risk of collision, each shall alter her course to starboard, so that each may pass on the port side of the other." Captain Calamai had two justifications for going against this well-known rule: he believed that the two ships were clearly on a starboard-to-starboard course, and he thought that an alteration of course to starboard might force his ship into the shoal water to the north of the light vessel.

Captain Calamai gave the fatal order, "Hard-a-port!"

When the two ships were less than five miles apart, the *Doria* still could not see the *Stockholm;* the *Stockholm* could see the masthead lights on the *Doria,* but not her sidelights denoting port and starboard. When the ships were about two miles apart—and at their combined speed of 40 knots, they would close in on each other in three minutes—the *Doria* could discern the first navigation lights on the *Stockholm* and could see that they were not green as expected in a starboard-to-starboard passing, but red. The red lights meant that the *Stockholm* was crossing the Doria's bow to make a port-to-port passing. At that moment *Doria's* Captain Calamai gave the fatal order, "Hard-a-port." His order was consistent with his plan for a starboard-to-starboard crossing. Then the crew watched helplessly as the bow of the *Stockholm* plowed into their starboard side.

The weather encountered by the *Stockholm* as it traveled east toward the *Nantucket Lightship* was very different from that encountered by the *Andrea Doria.* Somewhat before 11:00 P.M., as the *Stockholm* traveled through a clear moonlit and starlit night, third Officer Carstens-Johannsen first saw the sign of an approaching ship at the 12-mile radius on the radar screen. He recorded the position of the approaching ship as 2° to port of his heading marker.

Another plot of the ship a short time later also showed the ship to be slightly to port. As there was no sign of fog, Carstens, as he was called, saw no reason to call Captain Nordensen up to the bridge. But he did post a lookout on the port wing of the bridge to spot the approaching vessel, at which time he planned to alter his course to starboard to make a safe port-to-port passing. When the bridge lookout suddenly called out "Lights to port," Carstens immediately ordered a change in course of 20° to starboard.

Officer on the *Stockholm* watches in horror as the *Doria* is about to cross her bow

Then he went out onto the bridge and saw to his horror that an immense liner was about to cross his bow, presenting the green lights of her starboard side. He shouted "Hard-a-starboard!" to the helmsman. As Captain Calamai tried at the last moment to increase the distance in a starboard-to-starboard passing, Third Officer Carstens-Johannsen tried to increase the distance in a port-to-port passing. Then Carstens gave the orders "Stop" and "Full Astern."

The bow of the *Stockholm,* which was stiffened for travel through the ice in northern oceans, plowed into the side of the *Andrea Doria* just below the bridge down to her starboard fuel tanks with a momentum of 30 million foot-pounds.

The *Doria* listed because the starboard tanks were empty

Because only one main compartment of the *Andrea Doria* was damaged, she could have withstood the collision except for the severe list she developed soon afterward. Since the *Doria* was near the end of her journey, the starboard fuel tanks contained only air. When the starboard tanks were rent in the collision, water rushed to fill them, making the starboard side much heavier and causing a list to starboard of 18° and soon after, 22°.

The ship's list could have been corrected by making the port-side tanks as heavy as the starboard tanks. There was a narrow tunnelway leading to a pump room, which contained the valves for flooding the tanks with sea water. This tunnelway ran between the fuel tanks on the port and starboard sides, but since this tunnel was so far inboard, a watertight door had not been fitted to it. The tunnel flooded quickly after the collision. Thus no one could reach the valves to flood the port tanks.

Nearby ships carried off historic rescue

The severe list of the *Doria* had an unforeseen effect on the lifeboats. When the port-side lifeboats were readied, it turned out that they were useless. Their davits—the cranes that project over the sides of a ship—worked only if the vessel's list was less than 15°. The starboard-side lifeboats, which could not accommodate everyone, were also affected: they were hard to climb into because, with the list, they swung far out over the water.

A number of nearby ships rushed to the rescue, working against time as the *Doria*'s list gradually increased, making it more and more likely to capsize. Of the 1,706 passengers and crew on the *Doria*, 1,663 survivors were taken off. Most of the dead had been killed on impact. The surviving passengers were all off by 4:00 A.M. on July 26. At 5:30, when the list reached 40°, the officers also departed. At 10:09 A.M., 11 hours after the collision, the *Andrea Doria* rolled over onto her side.

Impact

In the legal proceedings that followed, each cruise line claimed that the other line was totally to blame and that its own officers were entirely blameless. The Swedish-American Line sued the Italian Line for $4 million; the Italian Line sued the Swedish-American Line for $1.8 million.

Stockholm has stronger legal case

The Swedish-American Line's case was by far the stronger. Most important was the immediate 18° list of the *Andrea Doria*, which put the responsibility for the capsizing and sinking onto the owners of the *Doria*, as did the *Doria's* turn to port to make a starboard-to-starboard crossing, thus violating the Rules of the Road. Also at issue were the speed of the *Andrea Doria* in fog, the failure of its crew to plot the bearings of the approaching *Stockholm*, and the disappearance of the *Doria's* ship's log.

The elements of the Italian Line's case against the *Stockholm* were third Officer Carstens-Johannsen's delay before making a starboard turn, his youth and inexperience, and the fact that the *Stockholm* was far to the north of the recommended track for eastbound vessels.

The ocean liners settle out of court

After the hearings had gone on for more than 3 months, the two lines settled out of court and dropped their claims against one another. This meant that the court did not apportion blame for the collision. It seems likely, however, that the two ships were not on safe parallel courses in which they would pass either port-to-port or starboard-to-starboard but were converging at a very small angle. A lesson relearned in the disaster is that in order to conform to the collision rules, officers of the watch:

- Must treat fine, or slight, bearings as head-on collisions
- Must make large alterations clearly discernible to the approaching ship in ample time

Officers must now take courses in radar

There have been considerable changes in both officers' understanding of radar and in the technology of radar since the *Andrea Doria* collided with the *Stockholm*. Captain Calamai probably did not understand that a 4° change in course would not make clear to the *Stockholm* his intent to make a starboard-to-starboard passing. However, today, largely as a result of this disaster, officers on ships are required to take courses in radar. In addition, since the 1980s international convention has required automatic radar plotting aids to be fitted in all large ships. Apparently the *Andrea Doria* made no plot of the *Stockholm's* bearings, and although the *Stockholm* presumably plotted the *Doria's* bearings, the plotting marks could not be produced, presumably because they had been rubbed out.

The true motion radars in use today allow the officer of the watch to see the actual course and speed of another ship, instead of the bearing relative to a heading marker. The latter system could result in errors because a ship can get a few degrees off course at any time. This may explain why the *Andrea Doria* thought the *Stockholm* was fine (slightly) to starboard and the *Stockholm* thought the *Andrea Doria* was fine to port.

Ships are now designed with separate ballast tanks

Another lesson learned from the collision concerned the ballast tanks. For reasons of stability, ships were permitted to list no more than 7°. Designers of the time, including those of *Andrea Doria*, satisfied the 7° limit by stipulating that the oil tanks, when empty, be flooded with sea water. However, the engineers did not want to contaminate their fuel tanks with sea water. In order to clean the tanks that had been flooded with sea water, the contaminated oil and water would have to be pumped onto barges for removal ashore.

After the sinking of the *Doria*, ship designers and engineers learned the advisability of having separate tanks for ballast water to provide stability. They also realized the importance of being able to remotely control the counterflooding valves for the ballast tanks.

Giving international weight to the new design direction of separate ballast tanks, the 1960 Safety of Life at Sea (SOLAS) convention criticized the practice of flooding oil tanks to provide ballast. SOLAS also recommended that full stability information be available to the ship's staff and that diagrams of ballasting arrangements be displayed on the ship's board.

New traffic separation scheme is introduced

A final effect of the *Andrea Doria–Stockholm* collision was the 1977 introduction of a traffic separation scheme for ships passing near the *Nantucket Lightship*. It required that eastbound ships keep to a lane south of the light vessel, and that westbound ships keep to a lane north of the light vessel. There is a separation zone three miles wide, with the light vessel in the middle, between the two lanes.

Where to Learn More

Barnaby, K. C. *Some Ship Disasters and Their Causes.* San Diego, CA: A. S. Barnes, 1970.

Hoffer, William. *Saved: The Story of the "Andrea Doria"—The Greatest Sea Rescue in History.* New York: Summit Books, 1979.

Marriott, John. *Disaster at Sea.* Ian Allan, 1987.

Moscow, Alvin. *Collision Course.* New York: Grosset & Dunlap, 1981.

Exxon *Valdez* runs aground

Bligh Reef in Prince William Sound, Alaska
March 24, 1989

Background

In the early morning of March 24, 1989, the tanker Exxon *Valdez* ran aground on Bligh Reef in Alaska, spilling 10.8 million gallons of crude oil into Prince William Sound. It was the worst oil spill in U.S. history. More than 1,500 miles of shoreline were polluted, and many thousands of birds and sea otters were killed. Although the immediate cause of the grounding was human error, the accident can be considered a technological failure, because oil is integral to how our technology-based society functions and because technology played an important role in the spill's aftermath.

An oil tanker runs aground Alaska's Prince William Sound and spills 10.8 million gallons of crude oil.

Many spills occur but few get wide news coverage

The Exxon *Valdez* oil spill and a few other spills have been widely reported, but there have been many other major spills the public has been generally unaware of. Worldwide between 1978 and 1990 there were 1,900 separate incidents of spills in excess of 10,000 gallons. Of these, 26 were of more than 10 million gallons, with the *Valdez* spill being toward the low end of this group. The Exxon *Valdez* spill became so widely known not only because it was the largest spill in the United States and in North American waters, but also because it did so much damage in the enclosed waters of Prince William Sound, where the oil could not disperse and break down as easily as it would have on the high seas. Before the spill, Prince William Sound—with its snow-capped mountains, many islands, and abundant wildlife—was a place of rare pristine beauty.

The Oil Pollution Act of 1990 requires all oil-carrying vessels operating in U.S. waters to be equipped with double hulls by the year 2015. Such a hull on the *Valdez* would have cut the spill by 60 percent.

When the Mideast cut off oil supplies, Congress authorized the Trans-Alaska Pipeline

The oil carried by the *Valdez* had traveled 800 miles by the Trans-Alaska Pipeline (TAP) from Prudhoe Bay to the port of Valdez. Alyeska, a consortium of oil companies that included Exxon, applied to Congress in 1970 for a permit to build a pipeline to transport the 9.6 billion barrels of oil that lay below the frozen ground of Alaska's North Slope. In 1973, with shipments of oil cut off by the countries in the Mideast, Congress voted to authorize the pipeline. Construction began in 1974 and was completed in 1977. Two million gallons of oil now travel through the pipeline every day, and 70 to 75 tankers leave the port of Valdez every month.

The environmentalists and Alaskan citizens who opposed building the pipeline feared that a catastrophic oil spill would be inevitable. To appease them, the federal government and the oil industry made promises, many of which were not upheld. In the early 1970s the federal government required double bottoms on tankers, and oil industry executives promised to build them. This issue was a key factor in congressional approval of the TAP route. But when the oil industry encountered hard times in the 1980s, double bottoms were mostly abandoned. Oil industry and federal officials also assured the public that ice conditions in Prince William Sound would be carefully monitored. They said state-of-the-art equipment would minimize chances that a tanker would collide with an iceberg. But this monitoring never occurred.

Safety standards decline in the 1980s

Safety standards also declined over the years. In 1977 the average tanker going out of Valdez carried about 40 crew members, but by 1989 the same size tankers carried half that number. Crews were also routinely working 12- to 14-hour workdays, and exhaustion was the norm. In the first years of the pipeline, Alyeska had a spill-readiness team that worked around the clock running drills and maintaining equipment. However, by 1982 public watchfulness slacked off. Spill-readiness team members were given other duties, and the team and equipment could not be as quickly assembled. The Coast Guard also experienced cutbacks in the 1980s. Fewer people were assigned to the Valdez Coast Guard station, and tankers were no longer inspected when they left the terminal. The radar at the Valdez station was replaced by a less powerful unit, which meant that tankers could no longer be tracked as far as Bligh Reef. The tanker crews knew nothing about the limits of the new radar. Had the old radar still been in use, the Coast Guard might have been able to warn the Exxon *Valdez* that she should turn sooner to avoid Bligh Reef.

Details of the Spill

The Exxon *Valdez* left the dock at Valdez at 9:12 P.M. on March 23, 1989, to embark on her five-day run to Long Beach, California. There had been a report of a heavy flow of Columbia Glacier ice into Prince William Sound, and this ice soon showed up on the ship's radar. A couple of hours into the voyage, the *Valdez* notified the Valdez Coast Guard station that she intended to leave the lane occupied by outbound vessels in the traffic separation scheme to use the lane for inbound vessels, hoping thus to avoid the ice.

However, the *Valdez* went farther out of her way than she had reported, crossing first the 2,000-yard zone separating traffic, then the 1,500-yard inbound lane, which also appeared to be full of ice, and continuing straight without making a right turn.

Bligh Reef lay six miles ahead; there appeared to be a gap of only nine-tenths of a mile between the edge of the ice and Bligh Reef, which meant that a well-timed turn would be critical. Six-tenths of a mile was needed to make the turn. Because the tanker was two-tenths of a mile long, the turn would have to start before the gap between the ice and the reef. The captain of the Exxon *Valdez*, Joseph Hazelwood, had 19 years experience with Exxon Shipping, but he was leaving the deck to go to his cabin. He told Third Mate Gregory Cousins to make a right turn when the vessel was across from Busby Island light and to skirt the edge of the ice, but Hazelwood did not specify an exact course.

Hazelwood guilty on only one charge

The tanker slipped past the light without starting to turn. Perhaps this happened because Cousins was looking at the radar screen. In any event, at 12:04 A.M. on March 24, 1989, the Exxon *Valdez* grounded on Bligh Reef. The bottom of the tanker tore open and spewed oil into the surrounding waters. Hazelwood was arrested. There was significant evidence that the captain had been drinking before boarding the *Valdez* and may even have been drinking during the voyage, but a jury subsequently acquitted him of all charges except negligent discharge of oil.

Response to spill is "slow and confused"

Spill management and cleanup were taken over from Alyeska by Exxon. Exxon was monitored by the EPA (Environmental Protection Agency), which represented the state, and the Coast Guard, which represented the federal government. Special interest groups also maintained a high profile during the cleanup. The response to the spill was, by all accounts, slow and confused.

Mobilizing the necessary equipment and materials was the first major challenge. It took 14 hours for the first barge to be loaded in Valdez and travel to the spill with containment booms, which are intended to float on the water and keep the oil from spreading. A day and a half into the spill, only a small line of booms were visible from the air. In addition to being too late and too few, the booms posed further problems. Some sank because of leakage or punctures in their buoyancy compartments, and some, which were intended to be used in protected harbors, were mistakenly deployed in open seas. However, about five days into the spill, after

The first major challenge in cleaning up the area was mobilizing equipment and materials. Containment booms, such as those shown here, took 14 hours to arrive at the spill.

the effort to contain the great mass of the spill had failed, the booms were used successfully to protect a number of fish hatcheries.

Prevailing currents carry oil 470 miles from spill

The lightering operation, when the remaining oil from the Exxon *Valdez* was transferred into another tanker, also worked fairly well. By early Saturday morning, more than 24 hours into the spill, the Exxon *Baton Rouge* was tied up alongside, and about four-fifths of the cargo of oil was recovered. Even so, by day four, the prevailing currents had carried oil 40 miles past islands and shoreline to the southwest; by day seven, 90 miles; by day eleven, 140 miles; and by day fifty-six, 470 miles.

Cleanup chiefs are reluctant to use high-tech remedies

Two other remedies—chemical dispersants and bioremediation—had unknown long-term effects, so these were used only minimally. Dispersants do not actually remove oil from the water; instead they break oil into tiny droplets that descend from the surface down through the water and, through rapid release of hydrocarbons from the oil, cause a severe toxic hit on marine organisms for about six hours. While it is generally considered better not to use dispersants close to shore where marine life is densest, some disagreement exists. Three days into the spill the Coast Guard authorized Exxon to use dispersants, but by that time gale force winds were blowing up, churning the oil into a mousse that could no longer be treated with dispersants. However, even if authorization had been secured earlier, there was not nearly enough dispersant available to treat a spill of this magnitude. The skimming equipment was also useless in rough seas, because mousse does not stick to the absorbent belts used to suck up the oil. By Monday morning, a blizzard was producing 25-foot waves, and less than 1 percent of the spilled oil had been recovered.

Bioremediation is the other controversial process. In this method, fertilizer is sprayed onto beaches to stimulate growth of naturally occurring bacteria, which degrade oil and change it to a harmless substance. Only 70 miles of shoreline were treated with this fertilizer, however, because of the uncertain long-term effects of the bacteria on other life forms.

Workers concentrate on cleaning shorelines

When efforts to contain and recover the spilled oil failed, the operation devolved into one of cleaning shorelines. At the peak of operations, more than 11,000 people, 14,000 vessels, and 85 aircraft were involved in the cleanup. The rocks along shorelines were washed with hot pressurized absorbent towels, or simply with absorbent towels. Many of the workers devoted their efforts to the wildlife—collecting dead animals so that the ingested oil would not get into the food chain, or bringing sick ones in for treatment. More than 100 endangered bald eagles, 36,000 seabirds, and 7,000 sea otters were found dead. Many more are thought to have died but were never found.

Oil damages birds first by impairing their water-repellency and their feathered insulation. Ingesting oil lowers their resistance to disease, reduces reproductive success, inhibits normal growth and development, and interrupts normal biochemical processes and behavioral patterns. Oil ingestion is harmful to organisms throughout the food chain.

Thirty-six thousand seabirds and more than 100 endangered bald eagles were found dead. Many more are thought to have died but were never found.

Of 43,000 lost wildlife, about 627 birds and 200 otters were saved in the rescue effort. Since each otter saved is estimated to have cost Exxon between $40,000 and $89,000, the rescue seemed little more than a symbolic or public relations achievement. It is also difficult to calculate the value from a strictly biological viewpoint. Scientists do not know how long it will take for populations to reestablish themselves, but they estimate that many species will need 5 to 20 years.

In September 1989 Exxon claimed to be finished with the bulk of its cleanup. Some of the shorelines appeared to be clean, but a layer of tar still existed not far below the shore surface. In some places the oil had penetrated four feet and threatened to penetrate deeper.

Impact

As a result of this catastrophic spill, Congress passed the Oil Pollution

Act of 1990, requiring all oil-carrying vessels operating in U.S. waters to be equipped with double hulls by the year 2015. The Coast Guard conducted a study after the spill that indicated that a double hull on the *Valdez* could have cut the amount of spilled oil by 60 percent. Another Coast Guard report showed that the cost of a double-hulled tanker, spread over the ship's probable 20-year lifetime, would be as low as three cents per barrel of oil carried.

Double-hulled tankers again mandatory—but they won't always work

Double-hulled tankers are believed to be advantageous in a low-energy grounding, in which the outer hull may be punctured, but the inner hull, nestled six feet or more inside the outer, remains intact. In a high-energy grounding, however, a rock or reef can rip both hulls apart. Japanese shipbuilders are championing an intermediate-deck design that may work better in high-energy groundings: a deck divides the tanker into upper and lower compartments, with the lower section below the ocean surface. If the hull is punctured, water will run in instead of oil running out, except for a small amount of oil released as the hull rolls.

In another response to the spill, the petroleum industry contributed $7 million to create the Petroleum Industry Response Organization, which established five regional centers along the coasts of the United States for fighting oil spills. These are staffed 24 hours a day and increase the availability of equipment for removing and cleaning up oil spills.

Legislators order Breathalyzer testing of pilots

The Alaska state legislature also acted to require all tanker captains leaving port to take a breath alcohol test no more than one hour before boarding. In addition, tankers now must keep two pilots on the bridge until after they have passed Bligh Reef, and tankers must be accompanied by two tugboats while still in Prince William Sound. Tanker pilots are no longer allowed to change lanes in the traffic separation scheme.

The litigation after the Exxon *Valdez* oil spill was divided into three phases in federal court, and the Alaska state court has considered $100 million in compensatory damages as well. The first federal phase revolved around two questions: who had been at fault, and was there any negligence regarding the attempted cleanup. The state sued Exxon, Alyeska, and all the other owner companies, and Exxon countersued, charging that the state should pay much of the cleanup cost because it had interfered with the use of dispersants. This phase ended with Exxon being found guilty of recklessness for permitting a captain known to abuse alcohol to command a supertanker.

By resolution of the second phase, which ended in August 1994, Exxon was ordered to pay $286.8 million in compensatory damage claims (the commercial fishers filing suit had wanted $978 million). Exxon claims it had paid out $3.4 billion making amends. Punitive damages were resolved by the third phase, which ended in September 1994. Exxon was ordered to pay $5 billion to Alaskans harmed by the spill, estimated at between 12,000 and 14,000 people and including commercial fishers, Alaska natives, property owners, and coastal municipalities. (Many estimated that the damages awarded would be as high as $15 billion; $5 billion is roughly Exxon's revenues for one year.) Joseph Hazelwood also was ordered to pay $5,000 in punitive damages for his role in the accident. Both Hazelwood and Exxon plan to appeal the verdicts.

Where to Learn More

Cahill, Richard A. *Disasters at Sea: "Titanic" to Exxon "Valdez".* London: Century, 1990.

Davidson, Art. *In the Wake of the Exxon "Valdez": The Devastating Impact of the Alaska Oil Spill.* San Francisco, CA: Sierra Club Books, 1990.

Keeble, John. *Out of the Channel: The Exxon "Valdez" Oil Spill in Prince William Sound.* New York: Harper-Collins, 1991.

National Research Council. Committee on Tank Vessel Design. *Tanker Spills: Prevention by Design.* Washington, DC: National Academy Press, 1991.

"Punishment for Exxon Is $5 Billion." *Detroit Free Press* (September 17, 1994): 1.

"Second Phase of Exxon Trial Focuses on Effects of Oil Spill." *New York Times* (July 10, 1994).

Skerrett, P. J. "Designing Better Tankers." *Technology Review* (September 1992): 8.

U.S. Coast Guard, American Petroleum Institute, U.S. Environmental Protection Agency (sponsors). *Proceedings of the 1991 International Oil Spill Conference.* San Diego CA: American Petroleum Institute, March 4–7, 1991.

"With 2 *Valdez* Oil-Spill Trials Concluded, the Big One Is Coming Up." *New York Times* (August 14, 1994).

U.S.S. *Squalus* sinks

Off Maine and New Hampshire coasts
May 23, 1939

A mechanical air-intake-valve operating gear fails during an underwater dive, allowing the aft part of the sub to fill with water and sink the submarine. Twenty-six crew members perish.

Background

On May 23, 1939, U.S. Navy submarine No. 192, the U.S.S. *Squalus*, sank in 240 feet of water off Portsmouth, New Hampshire. Twenty-six of the 59 crew members lost their lives. A large, hydraulically operated induction valve failed to close before the sub dived. The sea poured in and filled her aft portion and drove her to the bottom. Although the crew in the aft section of the sub perished, the 33 men who were rescued represented a triumph for the newly invented McCann Rescue Chamber. The rescue bell attached itself to the sub's hatch and enabled rescuers to extricate the men who were still alive but trapped behind the sub's closed, watertight doors.

New sub features latest design

The U.S.S. *Squalus* was one of the U.S. Navy's newest patrol-type submarines, which would distinguish themselves in the coming war with Germany and Japan. *Squalus* was first launched from the Portsmouth Navy Yard (in Kittery, Maine) in September 1938. A streamlined 310 feet long, 27 feet wide, and 450 tons in weight, she incorporated 25 years of submarine experience. Naval officials planned to test the innovative ship completely in spring 1939.

Squalus ran daily trials beginning May 16, 1939. The ship made 18 dives without incident and passed every test, meeting or exceeding all requirements. The required dives included a "by-the-numbers" dive, which was slow and careful, followed later by running dives. She also successfully performed quick dives and torpedo-firing dives. Official trials—necessary before a vessel can be accepted for service in the fleet—

The *Squalus* being raised at Portsmouth Navy Yard in September 1939. Repaired and recommissioned in May 1940 as the U.S.S. *Sailfish*, she sank seven enemy ships in the Pacific during World War II.

began May 23, 1939. Among that day's tests was the real-crash-dive simulation.

Sub's dives depended on exacting procedures

Squalus was to submerge 50 feet in 60 seconds. Before the dive, the sub would be at the surface, running on diesel engines. Before submerging, certain preparations had to be executed in correct sequence. *Squalus*'s main indicator board—or "Christmas tree"—would confirm proper execution with green "go" signals. When all crew members were below, the watertight hatches would be securely shut. Only the induction valves—large openings providing the diesel engines with the necessary air and allowing exhaust fumes to escape—remained open to the sea. Final dive preparation involved closing these valves. Only after the induction valves were closed would the green indicator light appear on the board. At this signal, *Squalus* would dive to periscope depth, then plunge quickly to the prescribed depth.

This important test procedure was performed off the Maine and New Hampshire coasts. *Squalus*'s personnel were experienced and highly qualified. Her skipper, Lieutenant Oliver F. Naquin, had already been commended by his admiral for excellent performance. The ship proceeded from the Portsmouth River into the sea, with 59 men aboard—the regular crew of 51 enlisted men and 5 officers, as well as 3 civilians from the Portsmouth Navy Yard.

Details of the Sinking

At 8:30 A.M. on May 23, 1939, Lt. Naquin ordered full speed for both diesel engines. The ship reached 16 knots (about 18 miles per hour), the prescribed speed to simulate a crash dive under attack.

The order to close the engine and hull induction valves was given. In the control room on the forward bulkhead, the second-class machinist's mate pushed the hydraulic manifold control handle upward through its 30 degrees of travel to the closed position. The control handle operates the hydraulic and mechanical connections to the large—31 inches across—manhole-cover valve. The indicator light on the board for the induction valve turned from red to green, and the Christmas tree board signaled green as well. This confirmed that the sub was watertight and ready to dive. The main ballast tanks were opened, allowing air to escape and water to rush in from the bottom. The sub dived, using the weight of the water she took on to make her heavy enough to submerge.

"The engine rooms are flooding!"

Squalus started her crash dive at 8:40 A.M. The sub was running on electric motors and plunged quickly, leveling off at a depth of 63 feet. Everything seemed routine, and the captain and crew were pleased with the results of the initial test. Congratulatory remarks were soon interrupted by a voice on the intercom, yelling: "Take her up! The inductions are open!"

Control room members had barely looked up when another, more frenzied voice screamed, "The engine rooms are flooding, Sir!"

The consequences of an open induction valve or ventilation duct are ominous. With these huge valves—2½ feet across—open, the ocean will rush through with the weight and pressure of all the water above. The engine room will be smashed by columns of water, filling it and the rest of the sub until she sinks. Knowing this, the captain gave immediate orders to blow ballast water in order to surface and then ordered all watertight doors closed to prevent the rest of the sub from filling up. With the order to surface, the engines of *Squalus* labored for about a minute, straining to overcome the 7½ tons of water that had been added to her weight. She hung motionless for a short time, but then her stern started to drop. She slowly slipped downward until she came to rest in the cold mud, 240 feet below the surface.

The captain was fully aware that by ordering the watertight doors shut, he was probably condemning those on the wrong side of the door to death—but his command was not only by the book, it was imperative if he were to save any of the crew. After these doors were closed, the storage cells, or batteries, had to be disconnected in order to save the lives of the crew. Unless the two disconnect switches are thrown immediately, the battery compartment will catch fire and explode. By the time the chief electrician's mate accomplished this dangerous task, the ship was sitting on the sea's bottom on an even keel with an 11 degree up-angle. The time was 8:45 A.M.

"Submarine sunk here. Telephone inside."

Captain Naquin then ordered a red smoke rocket fired and a forward marker buoy released. After the rocket rose in a smoky arc over the water, an orange, tin-can-shaped buoy floated to the surface. It was connected to the sub below by a telephone wire, and on its top were printed the words: "Submarine sunk here. Telephone inside." The captain ordered the firing of smoke rockets at 10:07, 10:24, 11:40, and 12:40. Amidst inventories of food, fresh water, and supplies of air, the 33 men still alive in the sub's forward torpedo and control rooms had no choice but to minimize all activity and sit and wait.

The navy began searching for *Squalus* at 9:30 A.M. when her surfacing report should have been received but was not. At 12:40 P.M. *Squalus*'s sister sub, the *Sculpin,* saw the red smoke rocket and headed toward her. At 12:55, 4 hours and 15 minutes after submerging, the crew sitting on the bottom were able to discern the beat of their sister ship's propellers. They listened as the anchor was dropped and the engines were cut, and finally the buoy phone rang. *Squalus* could only communicate the essential facts of her situation before the phone line snapped, but the link to the surface had been made!

Crew members are suffering CO$_2$ poisoning

As the men below waited silently, trying to keep warm and conserve oxygen, navy officials above decided quickly that the best chance of saving the crew was offered by the new rescue chamber method. Invented by Commander Allen R. McCann, the chamber would require much less time than trying to raise the sub. Cmdr. McCann was aboard the rescue vessel *Falcon,* which had steamed to the Maine coast for the rescue. The diver and the chamber needed time to be readied, however. When it was lowered at 10:00 A.M. the next day, conditions in the stranded sub had worsened. The temperature in the sub was a bitter 45 degrees and the air was increasingly toxic. Several members of the crew had begun to experience headaches, nausea, and vomiting—classic symptoms of CO$_2$ (carbon dioxide) poisoning.

At 10:15 A.M. the diver's weighted boots clanged as they struck the sub, and *Squalus*'s crew knew they were no longer alone. It took 25 minutes for the diver to attach to the submarine the all-important cable, down which the chamber would be lowered. Then the rescue chamber and its two-man crew began to descend. The rescue chamber was designed with a skirt to fit over the sub's escape hatch, out of which water could be blown.

The McCann invention worked as planned, and at 12:56 P.M. it began to rise to the surface, carrying 7 of the sickest crewmen. By 6:30 P.M. the rescuers had made two more trips, successfully rescuing 18 additional crew members. The fourth and final trip left at 7:30 P.M., but after picking up the final 8 men, the chamber became stuck 150 feet from the surface.

To remove the obstacle—a tangle in the cables—a diver was forced to cut one of the cables linking the chamber to the rescue ship at the surface. In order to gain buoyancy and allow the chamber to rise, therefore, ballast had to be released in careful increments. Maneuvering the chamber by ballast only meant it had no braking system and could easily shoot up through the bottom of the *Falcon.* With Cmdr. McCann giving instructions by phone, the chamber slowly and safely rose with the last eight of the *Squalus* survivors. It opened its hatch under the *Falcon*'s floodlights at 12:25 A.M.—nearly 40 hours after the submarine went down.

Impact

Just a few months after the *Squalus* sank, the U.S. Navy initiated its inquiry to learn the reasons for the tragedy. It discovered that the induction valve had failed to open properly some weeks before the accident but had always closed correctly. The valve had been removed and supposedly repaired before the mission on May 23. In addition to valve failure, the court found that the green signal light—the false confirmation of valve closure—was due to a malfunction in the electrical signal system.

Backup machinery too bulky for a combat sub

The inquiry also revealed that the navy did not incorporate all possible safety measures in the design of the *Squalus*. The navy had considered interlocking the sub's air valve and ballast tanks, which would prevent the sub from taking on ballast water to dive if the valves were open. But the machinery necessary to do this was considered too bulky for a combat submarine like *Squalus*. One board member asked if training the crew to close secondary valves as a habitual precaution might have lessened or prevented the disastrous sinking. When the board of inquiry eventually concluded its investigation, it:

- Blamed the sinking on the mechanical failure of the motor to close the induction valve

- Recommended that certain technical improvements be installed in the nation's fleet of submarines to avert further disasters

The ruling absolved the crew of any blame whatsoever.

McCann chamber proves its worth as rescue device

Another impact of the sinking was that the McCann chamber was proven to be a viable rescue method under certain conditions. The *Squalus* sank in a fortuitous situation: there were no appreciable tides to affect the diver or the chamber; she landed on the bottom mostly upright and horizontal; she sank near the base, was missed immediately, and was located quickly; and she sank in generally good weather. Despite the exceptionally good conditions, the men of the *Squalus* were in very bad shape by the time they were rescued and could not have endured much longer. No submarine sinking since has been as lucky.

Nearly four months after her sinking, the *Squalus* was raised, repaired, and given a complete overhaul. Recommissioned in May 1940

In addition to valve failure, the court found that the green signal light—the false confirmation of valve closure—was due to a malfunction in the electrical signal system.

with a new name—U.S.S. *Sailfish*—she went on to fight in World War II in the Pacific, where she sank seven enemy ships.

Where to Learn More

Barrows, Nathaniel A. *Blow All Ballast! The Story of the "Squalus."* New York: Dodd, 1940.

Lockwood, Charles A., and Hans Christian Adamson. *Hell at Fifty Fathoms.* Radnor, PA: Chilton, 1962, 203–227.

"Navy Affirms Finding *Squaius* Gear Failed." *New York Times* (February 2, 1940): 9.

Park, Edwards. "The Death Dive and Brave Rescue of the *Squalus.*" *Smithsonian* (January 1986): 102–210.

Shelford, W. O. *Subsunk: The Story of Submarine Escape.* New York: Doubleday, 1960, 70–81.

Stafford, Edward Peary. *The Far and the Deep.* New York: Putnam, 1967, 107–134.

"Whole Truth: Inquiry into the Sinking of the *Squalus.*" *Time* (July 3, 1939): 12.

U.S.S. *Thresher* sinks

Atlantic Ocean, 220 miles east of Boston
April 10, 1963

Background

On April 10, 1963, an advanced American nuclear-powered submarine, the U.S.S. *Thresher,* attempted to dive to its maximum operating depth of 10,000 feet. At 9:13 A.M., communication from *Thresher* to the accompanying surface vessel became erratic. At 9:17 A.M., the surface vessel heard the words "test depth" and then a muted thud. The entire crew of 129 men perished—and the cause of the accident remains undetermined. Many experts believe that carelessness and rushed construction resulted in a pipe failure and possibly a power loss, leaving *Thresher* unable to prevent a descent to fatal depths.

Nuclear submarine plunges beneath its "crush depth" and kills entire crew of 129.

First of new class of subs is powered by nuclear energy

The U.S.S. *Thresher,* with a price tag of $45 million, was the first of an advanced class of attack nuclear-powered submarines. These were to be true submarines, designed to spend prolonged periods of time in the depths of the sea. *Thresher* was powered by a Westinghouse S5W nuclear reactor. The reactor powered a steam turbine that drove the submarine's single propeller. *Thresher* also had a small backup diesel-electric power plant. This class of submarine would eventually carry nuclear-tipped underwater-to-underwater Subroc missiles.

Thresher featured an innovative tear-shaped hull. Though it made the sub very efficient underwater, the hull was primarily designed to accommodate the submarine's state-of-the-art sonar equipment, designated as BQQ-2. When installed in the specially shaped hull, the sonar equipment

The nuclear-powered U.S.S. *Thresher* before its 1963 sinking. The sub was 278 feet long, with a beam, or width, of 21 feet.

was protected from the interfering noise of the submarine's machinery and propeller.

Building began at the Portsmouth Naval Shipyard in New Hampshire. *Thresher*'s keel—the chief structural member that runs the entire length of the vessel—was laid on May 28, 1958, and the submarine was launched on July 9, 1960. Submerged, *Thresher* could achieve speeds of 30 knots or more, and she could go deeper than any previous submarine. At 278 feet long and with a beam of 21 feet, the sub displaced 3,700 tons of water.

Thresher can stay under for extremely long periods

Far greater underwater endurance was the most striking advance of the nuclear-powered submarine over her conventional, diesel-engine predecessors. A diesel-driven submarine stores energy in batteries, which drives electric motors to propel the ship. Energy storage capacity of batteries is limited, and diesel engines require oxygen. These factors mean that a conventional submarine can stay submerged for only a moderate period of time. Nuclear power, however, requires no oxygen and can produce oxygen for the crew through electrolysis of seawater.

Sea trials interrupted by equipment failures

During sea trials *Thresher* experienced a number of problems. In April 1961, when the submarine's trim (its buoyancy and ability to stay balanced) during fairly shallow dives appeared to be satisfactory, Commander Dean L. Axene ordered a deeper dive. As *Thresher* descended, her instruments showed that she was near the hull's pressure limit. Other instruments indicated that the sub was not even approaching her maximum operating depth. Axene refused to risk the crew, aborted the dive, and returned the vessel to the surface. When checked at the shipyard, the instruments proved to be faulty.

The submarine then resumed sea trials. Throughout this period *Thresher* would often make brief stops at port to conduct routine maintenance and to rest the crew. In November 1961 at one such stop in San Juan, Puerto Rico, *Thresher* underwent routine nuclear power shutdown and switched to diesel power. Eight hours later the diesel engines stalled, and the submarine switched to backup battery power. When extensive diesel repairs were found to be necessary, the nuclear system was reactivated, because backup battery power would not have been sufficient. However, the reactivation process would take several hours and could possibly overheat the reactor. During that time the crew had to work on the diesel engines in near-darkness and without ventilation. Temperatures inside the submarine reached 140°F before repairs were completed.

Thresher continued sea trials until she returned to the shipyard for Christmas. Trials resumed in February 1962 with *Thresher* undergoing shock tests to determine her ability to withstand underwater explosions. On June 3, off Cape Canaveral, Florida, a tugboat accidentally rammed the submarine and ripped a three-foot gash in her port side. After repairs *Thresher* endured two more weeks of pounding by underwater blasts. The sub suffered some damage but passed all the shock tests, then went into dry dock for a nine-month-long complete overhaul in Portsmouth.

Passing shallow test dives, sub heads for deep waters

On April 9, 1963, Lieutenant Commander John W. Harvey, the sub's new skipper, took *Thresher* out to sea. First they planned a series of shallow test dives. When these were completed, the submarine crossed the

Georges Bank area of the Atlantic to get beyond the continental shelf and into truly deep waters.

Details of the Disaster

On the morning of April 10, now in water between 7,800 feet and 8,500 feet deep and accompanied by submarine rescue ship *Skylark*, *Thresher* began a depth dive. *Skylark* could attempt a crew rescue in shallower, coastal waters. In these deeper waters off the continental shelf, however, *Skylark* could provide only a communication link and navigation checks.

Disaster strikes!

At 7:52 A.M. *Thresher* was 400 feet down. By 8:53 she approached her test depth of nearly 10,000 feet. Communication between the two vessels was normal up to this point. Starting at 9:13 A.M., however, communication from *Thresher* became garbled and sporadic. Then, sometime between 9:14 and 9:17, disaster struck. Communications personnel on *Skylark* heard the words "attempting to blow" and—just prior to a muted, dull thud—the words "test depth."

For nearly two hours thereafter, *Skylark* circled the area, continuously attempting to communicate with *Thresher*. Every ten minutes *Skylark* dropped hand grenades, the signal for a sub to surface. There was no response.

Rescue ship finds floating debris and an oil slick

Before dusk another submarine rescue ship sent to the area spotted an oil slick and a few bits of cork and yellow plastic. Within a few hours of searching, the navy had to conclude that *Thresher* must have exceeded its design depth and imploded. The Chief of Naval Operations notified the crew's families, scheduled a press conference to announce the loss, and prepared to mount the greatest search in U.S. naval history.

Impact

The navy employed scores of vessels to try to locate the lost submarine, including its new research ship *Atlantis* and the bathyscaphe *Trieste*.

The wreckage of the *Thresher* was photographed in 1964. The hull number, 593, is partially visible at right, slightly above center.

The bathyscaphe is a submersible—a vertical-movement vehicle with minimal maneuverability. *Trieste* had achieved the deepest dive on record and was the only vessel capable of going as deep as *Thresher*'s wreckage was thought to be.

Submersible vessel discovers *Thresher*'s wreckage

The trailing cameras of *Atlantis* photographed a mass of debris over an area approximately 1,000 yards by 4,000 yards. *Trieste* made its first dive to that spot on June 24, 1963. Several dives followed over the next two months until, on August 29, *Trieste* spotted a large quantity of fairly large pieces of metal. Using its newly installed mechanical arm, it picked up a piece of twisted, mangled brass pipe. The pipe carried its part number, job-order number, and the number 593, the numerical designation of the *Thresher*. This was the positive identification they needed.

The Naval Court of Inquiry had to make an official determination on how the tragedy happened. Investigators heard 120 witnesses. They ruled out enemy involvement. They also ruled it "inconceivable" that human error might have caused the disaster. The case file grew to 1,700 pages of testimony.

Sub would have imploded

The court concluded that the "most likely" cause for *Thresher*'s loss was a cracked fitting or a faulty joint in a cooling pipe. Such a break would fill a small compartment with seawater in seconds, possibly shorting out a key control panel or power circuit. Then the nuclear reactor would shut down, causing the sub to lose propulsion power. As the ship filled with water, her auxiliary diesel power would not be able to counteract the plunge. The tremendous counterpressure at these depths would disable the system that is supposed to expel seawater ballast. Under the increasing water pressure, *Thresher*'s hull would have twisted and rippled like rubber. Then she would have imploded (collapsed inward).

Alternate explanations

Although most experts tend to agree with the court's conclusion, two additional explanations have been proposed:

- An open sea valve may have been to blame. Openings in *Thresher*'s hull allowed water to be taken in for the nuclear power cooling system. If the submarine's electrical system had shut down because of a leak or any other reason, the valve, stuck in an open position, would have let seawater in continuously.

- *Thresher* may have encountered a mammoth underwater wave. A large storm moved across the Gulf of Maine on April 8, 1963. This storm could have generated a subsurface eddy, which is capable of generating an underwater wave 300 feet high. Such a subsurface wave could have driven the submarine far below its crush depth.

Unsettling facts emerge

The Naval Court of Inquiry brought to light a number of unsettling facts about *Thresher*'s rushed construction history. The periscope control had been installed backwards. A spike was found in the air circulation system. The labels indicating whether valves were open or closed were reversed on at least 20 percent of the sub's hydraulic valves. And tests on 145 of the 3,000 pipe joints revealed that 14 percent were below standard. Yet *Thresher* went to sea with 2,855 other joints untested.

Although investigators ruled out human error as the source of the disaster, the construction and dry dock records led them to state that "practices, conditions and standards existing at the time were short of those required to insure safe operation." The exact cause of *Thresher*'s loss remains undetermined, but it may trace back to this carelessness. The court also suggested that the navy began using the new design technology before its people had mastered how to use it.

In the wake of losing *Thresher*, the navy reformed some of its policies:

- A new quality-assurance program was put in place at naval shipyards.

- Workers responsible for soldering must fill out and sign a card verifying that each joint was properly soldered.

- Ultrasound replaced Xrays to inspect fittings and verify their integrity.

- Redundant hydraulic systems were installed that enabled valves to be closed without the use of electrical power.

- Submarines were fitted with a new emergency "blow" system that introduced air directly into ballast tanks at a rate seven times faster than on *Thresher*.

Technology advances in wake of disaster

The disaster also spurred technological advancement. The navy gave higher priority to developing highly maneuverable deep-sea submersibles. This resulted in the DSRV-1 (Deep Submergence Rescue Vehicle), which can rescue submarine crews down to 3,500 feet; the DSRV-2, which can operate at depths to 5,000 feet; and the nuclear-powered NR-1, which can operate at depths to 3,000 feet and can remain submerged for extended periods of time.

Navy declares investigation results "secret"

However, the navy earned considerable political criticism for the way it handled the *Thresher* disaster. It classified as secret all of the investigation's documentation—all twelve volumes—which meant that the navy's conclusion could not be evaluated independently.

Navy officials also appeared uncooperative during congressional hearings held by the Joint Committee on Atomic Energy. The committee members were justifiably concerned: a nuclear reactor now rests on the ocean floor 220 miles off the coast of Boston.

Where to Learn More

Bentley, John. *The "Thresher" Disaster.* New York: Doubleday, 1974.

Colley, David P. "The Lessons Learned from SSN 593." *Mechanical Engineering* (February 1987): 54–59.

Gannon, Robert. "What Really Happened to the *Thresher.*" *Popular Science* (February 1964): 102–109, 208–209.

Polmar, Norman. *Death of the "Thresher."* Radnor, PA: Chilton, 1964.

"*Thresher* Post-Mortem." *Scientific American* (November 1963): 66–68.

U.S.S. *Scorpion* lost

Mid-Atlantic Ocean
Reported missing, May 27, 1968

Background

On May 27, 1968, in the middle of the Atlantic Ocean, the submarine U.S.S. *Scorpion* disappeared with her crew of 99 without a trace. The navy launched one of the most extensive searches in its entire history to find the nuclear-powered submarine, but it located only portions of the ship on October 28, 1968, about 400 miles southwest of the Azores. The scanty evidence available prevented the Naval Court of Inquiry from pinpointing any clear cause to explain the *Scorpion*'s sinking. However, based on photographs of the wreckage, the court ruled out hostile action. The court blamed the accident on something internal to the submarine.

Unexplained sinking of a nuclear-powered submarine raises safety concerns and stirs development of deep-sea rescue and search submersibles.

Built to hunt down and destroy

The Electric Boat Division of General Dynamics Corporation built the nuclear-powered *Scorpion* (SSN-589) in Groton, Connecticut. An attack submarine of the Skipjack class, the *Scorpion*, armed with six torpedo tubes, was designed primarily to hunt down and destroy enemy submarines. She accommodated 10 officers and 89 crewmen, was over 251 feet long and 31 feet in beam, and displaced 3,075 tons of water.

The power to run this submarine came from a water-cooled Westinghouse S5W nuclear reactor with thermal power of 70 megawatts. With her Albacore-type hull shaped like an elongated teardrop, the nuclear-powered *Scorpion* achieved exceptionally high submerged speeds of 30 knots (about 35 miles per hour). Her propulsion plant made her the navy

The U.S.S. *Scorpion*, in a photo taken about eight years before she sank in the Atlantic Ocean. The sub was over 251 feet long, with a beam of 31 feet.

workhorse throughout the 1970s. There was also a lot of deliberate, built-in redundancy in the design of her major components—if one component failed, there was a duplicate to operate in its place. She had two steam generators, two pressurizers, two sets of turbines, and two turbo-generators. The sub also had conventional diesel engines and backup batteries.

Last voyage is to return home

The U.S.S. *Scorpion* was launched on December 19, 1959, and joined the Atlantic fleet in August 1960. She performed well during the next few years and was overhauled from June 1963 to May 1964. Following another tour of duty, she drydocked in February 1967 at the naval shipyard in Norfolk, Virginia, for another overhaul and refueling of her nuclear reactor. During October 1967 the *Scorpion* began sea trials under a new skipper, Commander Francis A. Slattery. After a tour in the Mediterranean in early 1968, the sub left Gibraltar to return to the Norfolk naval shipyard.

Details of the Disaster

On May 21, 1968, on course to the naval shipyard at Norfolk, the *Scorpion* reported her bearings south of the Azores and noted that everything was operating normally.

The U.S.S. *Scorpion* was never heard from again. No problem aboard the vessel was ever hinted at, and on May 27, 1968, she was reported overdue. On June 5, 1968, the chief of naval operations announced that the sub was presumed lost. The navy never learned any particulars about what actually happened on the *Scorpion* prior to or during her loss.

Impact

When it became apparent that the *Scorpion* was overdue, the navy launched 55 ships and 23 aircraft to search for the missing submarine and crew. Certain technical characteristics of the ship suggested to the navy where it should begin its search. When a Skipjack-class submarine goes down in the deep Atlantic, she descends past her crush depth—which is about 2,000 feet—and implodes (or collapses inward from the great exterior pressure). Navy antisubmarine warfare scientists knew that an event of this nature would be picked up by the navy's underwater listening system, called SOSUS. SOSUS consisted of an elaborate array of underwater devices that monitored the movements of foreign submarines.

Navy listening devices identify *Scorpion*'s location

They reviewed SOSUS tapes for the dates around which *Scorpion* vanished and discovered a peak noise event on at least two of their listening devices. This enabled them to get rough bearings on the location of the event. Then they wanted to test their theories about the noise on the SOSUS tapes and whether or not a nuclear sub in trouble would produce such sounds. Navy personnel went to the location where the SOSUS tapes recorded the sounds and detonated TNT charges. The amount of explosive was calibrated to create the same amount of noise, or acoustical energy, that would have resulted from the *Scorpion*'s implosion. The TNT noise matched the noise on the SOSUS tapes, so the Navy dispatched the *Mizar*, a naval research ship, to photograph the area.

Remote-control cameras photograph wreckage

The great depth involved made it impossible for divers to search the ocean floor. Instead the navy conducted sophisticated maritime detective work with trailing, remote-controlled 35-mm undersea cameras installed on the *Mizar*. Eventually experts identified wreckage of the submarine from photographs.

Several thousand undersea photographs provided views of the wreckage on the ocean floor. The bow appeared to have sunk partially into the sandy bottom. The submarine's "sail," or superstructure, was lying intact more than 100 feet away, broken away from the body of the *Scorpion*. The nuclear reactor was not visible.

Court of Inquiry pursues two theories

The naval Court of Inquiry began its work in June 1968. Before it received the *Mizar* photographs, the court was considering two hypotheses for the *Scorpion*'s loss:

- The submarine might have collided with a seamount (an underwater mountain).
- The submarine might have been victim of a hostile action—for example, a Soviet torpedo or depth charge—or she might have collided with another vessel.

Many people speculated that the *Scorpion* was on a secret mission and had been lost in waters frequented by Soviet submarines and surface vessels. During this period of the Cold War, hardly a week seemed to pass without some clash or minor incident between Soviet and American ships. In fact, a crew member had written home two weeks before the sinking that the *Scorpion* had been confronted by a Soviet destroyer with its guns trained on the American submarine.

Court rejects both theories

The seven-man Naval Court of Inquiry studied *Mizar*'s photos as well as other forms of testimony, meeting intermittently for 11 weeks and hearing from 90 witnesses. After evaluating these photographs, the court eliminated both its seamount collision and its hostile action/second vessel hypotheses. It issued its final report January 31, 1969, submitting this mainly nebulous conclusion: "The certain cause of the loss of the *Scorpion* cannot be ascertained from evidence now available." Although the court firmly discounted various potential causes, it was unable to develop any hypothesis to explain the *Scorpion* disaster.

Most observers who examined the *Mizar* photographs, however, believed that the origin of the *Scorpion* disaster was internal. As one assessment of the disaster maintained, "Had the *Scorpion* been hit by a torpedo or scraped by a surface ship while she was near the surface, this would have left identifiable damage."

Was the U.S.S. *Scorpion* unfit?

The court addressed another possibility: that the *Scorpion* might have been unfit or damaged prior to her disappearance. After all, the submarine was on her way back to the Norfolk shipyard. Testimony was submitted that the submarine had hairline cracks across her hull. Another witness stated that the submarine had a small oil leak around her propeller shaft. A crew member who was left ashore in Spain due to illness said that the *Scorpion* had been leaking hydraulic fluid around her periscope fitting and that she was having trouble with her navigational equipment.

The *Scorpion* did report at least five mechanical problems before leaving Spain. Moreover, the submarine had collided with a barge during a storm in the Naples harbor one month before she disappeared. This barge had been placed between the *Scorpion* and an American warship as a buffer during the storm, and the submarine's stern had taken the brunt of the impact. Divers made a partial inspection of the docked submarine but reported no significant damage. Finally, the navy's own Subsafe program had not been completely implemented on the *Scorpion*. This program, initiated in response to the *Thresher* disaster in 1963, involved a series of safety modifications. (For information about the *Thresher*, see the entry "U.S.S. *Thresher* sinks.")

Nuclear explosion ruled out

Even so, the court determined that the submarine's overall condition was "excellent" and that none of the pending repair work could have affected the *Scorpion*'s safety. It concluded that the disaster was not the result of the delayed Subsafe program. It also ruled out the possibility that her nuclear reactor had exploded, knowing that detection devices would have recorded such a catastrophe, just as they recorded the relatively contained implosion that did occur.

Because the *Scorpion*'s crew was well trained and prepared for emergencies, the court also rejected the idea that aberrant behavior on the part of the crew might have led to the disaster. They noted that there was no indication that any officer or seaman was unstable or unreliable. The court therefore discarded any prospect of the loss of the *Scorpion* being due to "the intent, fault, negligence or inefficiency of any person or persons in the naval service or connected therewith."

Navy court rejects other theories

Because the Court of Inquiry refused to speculate about possible internal failures, many outside observers complained that the court rejected such theories too readily. Among the causes they suggested were these three:

• A control failure

If the submarine were running very fast and very deep, her diving mechanism could have locked into the "dive" position. This would have sent the *Scorpion* below her crush depth before correction was possible.

• Leaks

Leaks were another area of concern, especially because—allegedly—there were cracks in the *Scorpion*'s hull and a faulty propeller shaft. Under the severe pressure of great depths, such flaws could lead to a sudden, fatal breach.

• Fire in the torpedo room

Malfunctioning of one of the submarine's own torpedoes—perhaps involving a fire in the torpedo room—might have caused such a disaster. The *Mizar* photographs did not rule this out, and some experts found this last hypothesis the most feasible.

Navy renews development of deep-sea rescue vehicles

Nevertheless, to date no exact cause for the *Scorpion* disaster has ever been determined. Losing the *Scorpion* and not being able to explain what happened to it generated enormous controversy for the U.S. Navy. Eventually the political storm subsided, but a logistical issue remained: the navy had no capability to rescue any submarine at depths below 1,300 feet. The navy began a research and development program aimed at producing deep-sea rescue vehicles after it lost the *Thresher* in 1963. Losing the *Scorpion* in 1968 gave new impetus to this program.

Navy develops DSRV-1, DSRV-2, and NR-1

The navy's first Deep Submergence Rescue Vehicle (DSRV), the DSRV-1, was launched in 1970. Shaped like a torpedo, this 50-foot vessel with a

The navy had no capability to rescue any submarine at depths below 1,300 feet. It began a research and development program aimed at producing deep-sea rescue vehicles after it lost the Thresher *in 1963. Losing the* Scorpion *in 1968 gave new impetus to this program.*

fiberglass hull could operate to depths of 3,500 feet. The DSRV-1 could link to the submarine's escape hatch and take 24 crew members onboard to carry to the surface. The DSRV-2, built by Lockheed, was able to operate as deep as 5,000 feet. Both submersibles could be transported by air or water or even carried on a nuclear submarine itself, which could launch and recover them without surfacing.

The navy also built an experimental rescue craft called the NR-1. About 136 feet long and with an operating depth of 3,000 feet, the NR-1 looked like a small submarine. It had room for a crew of 7 and was powered by a small pressurized-water reactor. Its two propellers and tunnel thrusters made it highly maneuverable. Among its special features were sonar, for navigating and locating sea-floor objects, and a large recovery device.

Although none of these rescue submersibles has ever aided sunken submarines, they have been used for comparable missions. In 1976, for example, the NR-1 helped rescue an F-14 fighter that rolled off a navy aircraft carrier in 1,960 feet of water.

The possibility of nuclear contamination from the *Scorpion*'s reactor raises a final issue. The submarine's debris is scattered over a large area. As her hull was compressed from both ends during the disaster, it is not possible to assess the condition of the reactor compartment. However, the seabed around the wreckage of the *Scorpion* has been investigated three times. Samples show minor amounts of cobalt-60, which indicate that there is some leakage from the primary reactor system. In March 1975 Admiral Hyman G. Rickover reported to Congress that although the cause of the *Scorpion* disaster remains unspecified, there is still "no evidence that it was due to a problem with the nuclear reactors."

Where to Learn More

Erikson, Viking O. *Sunken Nuclear Submarines: A Threat to the Environment?* Norwegian University Press, 1990.

"Finding the *Scorpion*." *Newsweek* (November 18, 1968): 104.

"Loss of *Scorpion* Baffles Inquiry." *New York Times* (February 1, 1969): 1.

Middleton, Drew. *Submarine: The Ultimate Naval Weapon; Its Past, Present, and Future.* New York: Playboy Press, 1976.

"Presumed Lost." *Newsweek* (June 17, 1968): 74.

"Sub Found—But Mystery Remains." *U.S. News & World Report* (November 11, 1968): 12.

11

Airships, Aircraft, and Spacecraft

U.S.S. *Shenandoah* crashes

Background

On the morning of September 3, 1925, a severe thunderstorm destroyed the first U.S.-built rigid airship over Ohio. The U.S.S. *Shenandoah* was heading west from Lakehurst, New Jersey, when it encountered a catastrophic storm. The powerless 680-foot airship was forced up and down and twisted like a rag doll until it broke apart. Of its crew of 43, 29 survived.

Fourteen people perish as a storm destroys the first American-built airship, which triggers reevaluation of airship design and criticism of military policies.

American airship will use helium

In 1919 the U.S. Secretary of the Navy authorized the construction of a rigid airship—a lighter-than-air craft. Airship number ZR-1, later to be christened the *Shenandoah*, was designed after a German zeppelin deployed during World War I. The French had captured the zeppelin after forcing it to land. American engineers modified the German design by strengthening the frame and increasing the airship's length by 33 feet. This extra length was necessary because helium, not hydrogen, would give the airship its lift. Although helium gas is completely safe—unlike flammable hydrogen—helium provides only 92.6 percent of hydrogen's lift. The first rigid airship to be built and flown by Americans would also be the first ever to rely on helium.

Construction on the airship began at the Naval Aircraft Factory in Philadelphia in 1920. The ship made its maiden flight at Lakehurst September 3, 1923, and one month later was christened the *Shenandoah*, an American Indian word meaning Daughter of the Stars. It was 680 feet long, 78.7 feet in diameter, and held 2.148 million cubic feet of helium gas.

The wreckage of the *Shenandoah* after it broke apart over Byesville, Ohio, and crashed. Fourteen of the crew of 43 died in the disaster.

Its five 300 hp (horsepower), six-cylinder Packard engines gave it a maximum speed of 60 mph, and its fuel tanks allowed it a range of about 3,000 miles.

Navy plans publicity trip to West Coast

Instead of the hangaring the airship as the Germans did, the Americans decided to moor their airships to masts. These tall, steel-framed towers would anchor the airborne ship by its nose, like a balloon tied down by a string. Opting for mooring masts eliminated the expensive prospect of building huge (about 1,000 feet by 300 feet) hangars and would enable any airport with a tall mast or tower to accommodate the airship.

Although the *Shenandoah* was experimental, the government ambitiously planned for it to make a historic flight over the North Pole and therefore erected a network of mooring masts across the West Coast and up into Alaska. When plans for this trip were canceled, the navy decided to exploit the public's enormous interest in the *Shenandoah* by scheduling a more modest publicity trip to the West Coast, using the network of masts for stopovers.

Airship fortuitously survives January gale

During January 1924—before the first extended publicity trip—the *Shenandoah* was blasted for four days by gale-force winds while anchored to its mast at Lakehurst. Twisted on its longitudinal axis, the rolling airship eventually ripped away from the mast, leaving its nose tied fast to the tower. With a combination of skill and luck, the crew of the damaged airship just managed to ride out the storm and coast back to Lakehurst. The fact that the *Shenandoah* survived the vicious weather was taken as evidence that the airship was structurally sound and ready for anything.

Helium presents special problems

By May 1924 the airship had been repaired. Under the direction of U.S. Navy Commander Zachery Lansdowne, the *Shenandoah* undertook a series of mooring trials at sea, attaching itself to the U.S.S. *Patoka*. Then the airship left for a transcontinental flight October 7, 1924. During this flight the problems peculiar to helium airships became apparent. First, there was a loss of performance compared to hydrogen, a flammable gas that provides greater lift. Second, because helium was scarce and expensive, it had to be conserved, which compelled various expedients to minimize its release.

The *Shenandoah* was kept at 85 percent inflation before flight, because the helium gas expanded as the airship rose. Its "pressure height"—the altitude at which inflation reached 100 percent—was 4,500 feet. The ship therefore had to remain below 4,500 feet to prevent the automatic valves from releasing helium. Landing also led to problems in helium conservation. When the Germans landed their zeppelins, hydrogen gas was simply expelled through valves until the airship lost buoyancy. Since routinely releasing large amounts of helium was unthinkable, the Americans decided they would land the *Shenandoah* at night, when the helium became colder and heavier.

Trans-U.S. trip takes risks to make history

For its first publicity flight across the United States, which was essentially a training trip, the *Shenandoah* could easily have crossed high above the Rocky Mountains. But this would have meant releasing thousands of cubic feet of precious helium. The airship therefore threaded its way through many treacherous mountain passes, risking collision with cliffs. Cmdr. Lansdowne successfully steered the *Shenandoah* through the Rockies on both legs of the trip and safely returned to Lakehurst October 25, 1924.

The *Shenandoah*'s trip was a scientific and engineering phenomenon and generated enormous public interest. Until this flight, transcontinental passenger service was unknown, and only mail had been flown from New York to San Francisco.

The *Shenandoah*, nonetheless, would not fly again until 1925. Its helium was temporarily transferred to the newly christened *Los Angeles* (airship number ZR-3), which had been purchased from Germany. Helium was so scarce that the United States could not supply two large airships at the same time.

Meanwhile Cmdr. Lansdowne took measures to further conserve helium in the *Shenandoah*. He had 8 of the ship's 16 maneuvering valves removed and gas-tight "jam pot" covers installed over the automatic valves. These modifications would reduce gas loss, but they also created the potential for another problem: if the ship were to ascend too high too quickly, it could break up due to excess pressure. Lansdowne successfully argued to the skeptical Bureau of Aeronautics, whose approval he had to secure, that, if necessary, the ship's maneuvering valves could be opened by hand.

Navy plans to show airship to Midwesterners

A battle Lansdowne did not win concerned the scheduling of the *Shenandoah*'s next publicity flight, this time to the Midwest in August. A native of Ohio, Lansdowne knew very well to expect unstable weather over this region in the summer. Lansdowne argued to have the trip postponed until fall, but the navy granted only a two-week delay, to the beginning of September. The navy then committed to this decision by publicizing the *Shenandoah*'s schedule to coincide with the many state fairs planned for that month. The navy wanted the *Shenandoah* to get maximum exposure to the hundreds of thousands of Midwesterners who had never seen an airship.

Details of the Crash

At midafternoon on September 2, 1925, the *Shenandoah* set off from Lakehurst on its publicity trip to the Midwest, its fifty-eighth flight. The crew's wives had come to watch the departure, and according to custom they turned their backs on the airship as it headed west—it was considered bad luck to watch as an airship faded out of sight. The *Shenandoah* was greeted at Philadelphia by car horns and factory whistles an hour and a half later, and it soon reached the Allegheny Mountains. At around 3:00 A.M. on September 3, the crew first observed flashes of lightning, initially behind them to the east and then ahead to the west and northwest. An hour later, at 2,100 feet over Byesville, Ohio, the airship ran into strong headwinds that all but stopped its forward progress. Minutes later, the airship began to rise.

Airship rolls and twists and breaks in two

Ominously the *Shenandoah* continued to ascend despite every countermeasure the crew attempted. The engines labored, but the airship had encountered a freak confluence of air currents in which a stream of south-moving cold air was overriding a mass of warmer air coming north. As the warm air rose, so did the airship, at a rate of 360 feet per minute. After briefly steadying at about 3,000 feet, the ship began to rise even faster, at a terrifying rate of nearly 1,000 feet per minute. As it became imperative to release helium, Cmdr. Lansdowne ordered his crew to open the manual valves and uncap the automatic valves.

When the *Shenandoah* stopped rising, at 6,200 feet, the crew prepared for the inevitable plunge. The airship began to plummet even faster than it had risen, and to slow its dive the crew released tons of ballast water. After this measure took effect, however, another ear-popping skyward ascent began. The *Shenandoah* was now rolling and twisting like a living thing. Fierce winds hammered its hull, forcing its nose to port and its tail to starboard. The crew attempted to release the fuel tanks to drop additional weight, but—spinning with its nose pointed 30 degrees upward—the airship broke in two.

As airship breaks up, 14 die

The *Shenandoah* split apart about one-third of the way from the front. The control car, with Cmdr. Lansdowne among a crew of eight, was catapulted away like a stone; none survived. Seconds later four more crewmen fell with the engine car. The huge bow (front) section, with its crew of seven, then started to rise rapidly, buoyed by some still-intact gas cells. This 225-foot bow section spun like a top for another two hours. All of its crew members, nonetheless, managed to survive by clinging to broken cables and girders until the bow section finally touched down 12 miles away from the ini-

The Court of Inquiry's conclusion triggered a reevaluation of airship design. To allow for severe aerodynamic loads, subsequent American airships were made much stronger, though doing so sacrificed lightness. Because all but two deaths in the Shenandoah came when the suspended control and engine cars broke off, newer ships incorporated these functions directly into the hull.

tial breakup. The rear section of the ship broke into two and slowly drifted to earth, dragging along the hills and leaving a trail of injured men.

Twenty-nine of the crew of 43 survived. In the aftermath of the disaster, curiosity seekers swarmed over the various crash sites and literally picked the wreckage clean. Souvenir hunters stripped everything possible from the ship's remaining parts. Some of the covering later appeared as raincoats sold as Shenandoah Slickers.

Impact

The Court of Inquiry investigating the disaster concluded that the *Shenandoah* had broken apart because it was not strong enough to withstand the strains imposed on it by vertical air currents. It left unanswered whether Cmdr. Lansdowne's valve-capping system resulted in or contributed to the breakup.

Disaster affects airship design, invites criticism of military

The court's conclusion triggered a reevaluation of airship design. To allow for severe aerodynamic loads (stresses), subsequent U.S. airships were made much stronger, which sacrificed lightness. Because all but two deaths in the *Shenandoah* came when the suspended control and engine cars broke off, newer ships incorporated these cars directly into the hull. The military's decision to use helium rather than hydrogen, however, was vindicated. Since a hydrogen fire would certainly have erupted during a breakup like the *Shenandoah*'s, the survivors owed their lives to the use of helium.

An additional result of the *Shenandoah* crash was outspoken criticism of the American military. Most significantly, U.S. Army Colonel Billy Mitchell issued a 17-page indictment of the military's role in the disaster. Mitchell claimed that this and other disasters were the direct result of incompetence and even criminal negligence. Mitchell was court-martialed, convicted, and suspended from rank, duty, and pay for five years for his criticism. Even so, Mitchell achieved his purpose: focusing national attention on the military's aviation policies and practice.

Where to Learn More

Clarke, Basil. *The History of Airships.* New York: St. Martin's, 1961.

Hook, Thom. *"Shenandoah" Saga.* Annapolis, MD: Airshow Publishers, 1973.

Jackson, Robert. *Airships in Peace and War.* London: Cassell, 1971.

Payne, Lee. *Lighter than Air: An Illustrated History of the Airship.* Orion, 1991.

"Technical Aspects of the Loss of the U.S.S. *Shenandoah.*" *Journal of the American Society of Naval Engineers* (August 1926): 489.

Walker, J. Bernard. "Tragedy of the *Shenandoah.*" *Scientific American* (November 1925): 301.

Wensyel, James W. *"Shenandoah." American History Illustrated* (February 1989): 24.

R-101 crashes

Beauvais, outside of Paris, France
October 5, 1930

Untested, overweight, unstable, and underpowered, a British airship battles severe weather over France. Forty-eight people perish in the fiery hydrogen explosion.

Background

In 1924 the British government initiated the Imperial Airship Scheme, the brainchild of Sir Dennistoun Burney, a member of Parliament. This program would prove once and for all whether very large, rigid airships were commercially feasible for long-distance, high-speed flights. The plan called for a fleet of six airships to link Britain with far-flung parts of its empire, to Australia, Canada, India, and South Africa. Carrying passengers and mail, airships promised to cut travel time. A voyage from Singapore to England by steamer took 24 days. By airship it would take only 8 days.

The original plan was cut back to two airships when the Labour party came to power. The government would build one ship, and private industry would build the other. The government's airship, R-101, came to be known as the "Socialist Ship" because of the left-leaning sympathies of the Labour party. The R-100 was called the "Capitalist Ship." Fierce competition between the two camps spurred a bitter rivalry.

Politicians make unfit advisers

Construction of the R-100 proceeded generally along normal lines, but development of the R-101 was subject to many nontechnical factors. The question of how to power the airship typified how politics influenced the government ship. Both design teams originally planned to use diesel engines, because diesel fuel was considered safer than gasoline. When both realized that diesels would be far too heavy, the builders of the R-100 simply switched to six proven Rolls-Royce Condor engines. The builders

The British airship R-101 was the largest, costliest, most complex airship ever flown. Its radical design featured internal rings that had no bracing wires between them.

of the R-101, however, felt they could not abandon diesel engines because they had publicly proclaimed their safety. Similar decisions were made regularly with the R-101. With so much publicity showered on the Socialist Ship, political leaders were reluctant to trash their technically bad plans.

Five years later and actually ahead of the R-100, the R-101 came out of its hangar. Though a truly remarkable sight, the airship displayed wonderful but inappropriate technology:

- Immense diesels
- Special servo-mechanisms (feedback systems) that controlled its rudders and elevators with power instead of by hand
- A complex system of specially designed, supersensitive valves that released gas with the slightest variation in pressure (such as when the ship pitched or rolled)
- A frame that included strong but heavy steel (the R-100 used only the much lighter Duralumin)
- Its most fundamental design departure: The use of internal rings with no bracing wires between them

The result was an incredibly heavy airship without much lifting ability.

Comfortable and luxurious, but airship can't go the (long) distance

R-101's maiden flight took place on October 14, 1929, and by the end of November it had logged seven trips totaling 70 hours of flight. All were in good weather; none was at high speed. Airborne, the ship was a true spectacle and even more so from inside. Its palatial passenger quarters included an asbestos-lined smoking room, a dining salon, a huge lounge with potted plants, and even a wide, picture-windowed promenade.

The comfort and luxury couldn't hide the fact that the ship staggered through its flights. Dangerously overweight and underpowered, it weighed 113 metric tons. Though its lifting capacity was 148 tons, its disposable, or actual, lift was only 35 tons—not enough to carry passengers on long-distance flights. Its frustrated designers returned it to the hangar to cut its weight.

Renovated R-101 is largest, costliest airship ever

From November 1929 to June 1930 five tons of expensive equipment, including the power-steering system, were ripped out of R-101. The ship was cut in half, and 45 feet of hull and an extra gas bag were added inside. On October 1, 1930, the R-101 emerged from the hangar an entirely new ship. It sported a new cover and boasted 49 metric tons of disposable lift. At 777 feet long, it was 2 feet longer than Germany's *Graf Zeppelin* and the largest, costliest, most complex airship ever flown.

But it had yet to pass its test finals.

When rival R-100 flew a successful round trip from England to Canada, Britain's secretary of state for air, Lord Thomson, felt compelled to fly the R-101 to India. The energetic secretary spoke often and publicly about flying to India for an important conference and returning to London by October 18 for another.

The engineers were not as enthusiastic as Thomson, because the rebuilt ship had not yet passed its test trials, nor had it received its certificate of airworthiness. The certificate was required to fly legally. As political pressure mounted for Thomson's flight to depart as planned, a conditional certificate was issued, stipulating that full-speed trials be carried out during the voyage. Lord Thomson declared that R-101 was "safe as a house, except for the millionth chance." The doomed ship departed at 6:36 P.M. on October 4, 1930.

Details of the Failure

In the frantic hours before the flight to India, many changes were made that added back much of the weight that had been removed during rebuilding. A pale blue carpet was installed to cover the entrance corridor and lounge. Six hundred feet long and the width of a tennis court, the carpet alone added 2 to 3 metric tons of weight. Crew members were limited to 10 pounds of personal baggage each, but Lord Thomson's baggage weighed in at more than 1 ton.

Crew tops off the fuel tanks

A planned state banquet required adding crates of champagne, barrels of beer, and boxes of silverware to the airship's load. And Lord Thomson's tight schedule did not allow any time for refueling at their first stop in Ismailia, Egypt. So the crew topped off R-101's tanks with extra fuel, an additional 9 tons to the load.

To compound these problems, a gale was rising over France in the very direction the ship was headed. Despite severe weather predictions,

The remains of the R-101 after it crashed in France. Officials disregarded predictions of severe weather and embarked on the doomed flight.

the airship lifted off at 6:36 P.M. on October 4, 1930, with 54 people aboard, 6 of whom were passengers. Instead of rising when the airship was released from the mast, it began to sink slightly. Crew members unloaded 4 tons of water ballast, and the airship rose sluggishly. While dropping ballast, or extra weight, is standard technique to compensate for expected gas loss during a flight, it is hardly standard operating procedure at the outset of a flight, just to get the ship to rise.

Airship explodes, killing 48

The ship lumbered across the English Channel, flying dangerously low on an undulating corkscrew path, rising and falling 400 to 500 feet at a time. As the R-101 flew over the French town of Beauvais, 40 miles northwest of Paris, it was pounded by winds of 50 miles per hour. Shortly after the ship's watch changed at 2:00 A.M., the aircraft suddenly went into a steep dive. The crew used the elevators (the movable horizontal surfaces at its tail) to stop its plunge and level the ship, but just as it seemed to stabilize, it pitched down a second time. At 2:09 A.M. on October 5, 1930, the R-101 fell to the ground at less than 5 miles per hour. It bounced lightly, skidded some 60 feet, then exploded into a fireball.

Tremendous explosion knocks witness to the ground

The crash was witnessed by a French rabbit trapper setting snares in the woods. The witness reported that the nose-down landing was immediately followed by "a tremendous explosion that knocked me down. Soon flames rose into the sky to a great height. Everything was enveloped by them. I saw human figures running about like madmen in the wreck. Then I lost my head and ran away." Only eight people managed to get clear of the inferno, and of these, only six survived. Lord Thomson and all the ship's officers died in the flames.

How airship's doom was sealed

A court of inquiry judged that one or more of the ship's forward gas bags deflated, causing the ship to go down, but investigators could never establish the cause of the accident officially beyond question. More than 50 years after the crash, however, Sir Peter G. Masefield produced a study, now regarded as definitive, describing 13 separate but cumulative factors that doomed the airship:

1. The India flight as a prototype trial: Committed to as early as 1924, the flight to India had to be accomplished at the earliest possible date.

2. Weak outer cover: Only part of the ship had received a new cover after the extra bay was installed, and the nose and stern portions of the gas bag were original.

3. No full-speed trial: The airship was never tested when fully stressed before it embarked to India.

4. Overconfidence and inadequate trials: These faults were believed to be common at the time and not unique to R-101.

5. Decision to embark: Despite an unfavorable weather report, officials still decided to go ahead with the trip.

6. Scott's judgment: Major G. H. Scott refused to admit that airships were not "all-weather" crafts. He permitted departure for India when most captains would have waited for better weather.

7. Engine starting delay: It took nearly 30 minutes to get the starboard forward engine working.

8. Weather: Although the report predicted heavy rain and low-level turbulence, conditions were actually much worse.

9. Fast cruise regime: The captain ran his engines at full speed throughout the flight. This subjected the unproven airship to stresses no one knew if it could handle.

10. Change of watch: At the most critical point in the flight, the more seasoned men were replaced by less experienced crew members.

11. Crew fatigue: The majority of the crew had been on continuous duty since R-101 left its shed on October 1, three days before embarking on the India flight.

12. Engines throttled back: This is an instinctive maneuver to slow down the ship, but employing it just as R-101 pitched downward diminished its lift and speed and contributed to its dive.

13. Calcium flares: Installed in the control car for use during the flight over the English Channel, the flares ignited instantly on contact with moisture from the released water ballast and precipitated the hydrogen explosion.

Sir Peter Masefield explained that each of these factors contributed to the disaster. However, he stressed that the fatal chain of events could have been broken had any one of them not occurred. In his estimation the predominant factors that caused the R-101 disaster were the weather, the weak outer cover, the decision to throttle back, and the calcium flares.

Impact

This disaster finished the airship enterprise for Britain. Shock over the disaster and the loss of Britain's elite airship corps affected the British people more than any crash since World War I. Fed up with ill-starred airships, the British public lamented the loss of the crew, "Killed to make an Air Minister's holiday." "Ban the Gas Bags" became a popular newspaper headline. As soon as the airship program was officially abandoned, even the successful $2 million R-100 received the death sentence. Smashed to pieces with axes, flattened by a steamroller, and sold as salvage, it brought in just $2,500.

Where to Learn More

Botting, Douglas. *The Giant Airships*. Alexandria, VA: Time-Life Books, 1980, 122–131.

"Britain's Dirigible Horror." *Literary Digest* (October 18, 1930): 10–11.

"British Airship R-101 Is Destroyed in Crash and Explosion in France." *New York Times* (October 5, 1930): 1.

Collier, Basil. *The Airship: A History*. New York: G. P. Putnam's Sons, 1974, 166–210.

Masefield, Peter G. *To Ride the Storm: The Story of the Airship "R.101."* William Kimber, 1982.

"Weight of Rain Forced Down R-101." *New York Times* (October 6, 1930): 1.

Hindenburg explodes

Lakehurst, New Jersey
May 6, 1937

A hydrogen fire destroys the last and largest commercial airship.

On May 6, 1937, the largest airship ever built, the German *Hindenburg*, crossed the Atlantic Ocean and dropped its mooring ropes to the ground crew below. Suddenly it exploded and burst into flames. Thirty-two seconds later the entire ship lay crumpled and smoldering on the ground. Of the 97 people aboard, 62 survived the fiery crash. Airship development, however, ended forever.

Background

Airships, or lighter-than-air crafts, improved the technology of balloons, which were invented in 1783. Balloons used bags made of airtight fabric. The inventors of balloons, Jacques and Joseph Montgolfier of France, discovered that when air is heated, it expands and weighs less than the same volume of cold air. Since hot air is lighter than plain cold air, it proved able to lift a balloon 1,000 feet off the ground.

This discovery stirred up great interest in ballooning, and less than three months later, J. A. C. Charles of France launched a balloon filled with hydrogen, a gas lighter than air. Within a year people succeeded at flying in balloons, but steering them was not possible—they went wherever the wind blew them.

Balloons become steerable

Steering was attained when propellers and rudders were added to the aircraft, and thus airships were born. Then other improvements appeared.

The *Hindenburg* flying over New York City; the airship made the transatlantic trip several times before it crashed.

The propellers were driven by a power source, such as steam, their shape was stretched horizontally, and some were given a kind of skeleton inside. Cars—known as gondolas—were attached underneath, providing a place for people to ride. These new airships were also called dirigible balloons, or dirigibles.

Dirigibles useful in war

A retired German army officer, Count Ferdinand von Zeppelin, had great success with his airship design. With a rigid frame inside made of aluminum, this airship traveled at 20 miles per hour. The German government began to invest in airship development and used them—as did other countries fighting in World War I—as war machines. They carried bombs and machine guns. They conducted patrols. They were also enormous targets, so their technology was forced to improve. Zeppelin speeds reached 80 miles per hour, and they could fly at 20,000 feet.

Hindenburg Establishes Transatlantic Air Service

When Germany lost World War I, German airship development ceased for a time, but the Zeppelin company, in partial payment for having supplied airships to the German war effort, was required to build two airships for the United States. These airships, the *Los Angeles* and the *Graf Zeppelin*, carried passengers, mail, and freight. Their successor was built in 1936. Designated the LZ-129, it was named the *Hindenburg*, and it established transatlantic air travel as a commercial reality.

The *Hindenburg*'s designer, Hugo Eckener, succeeded Count Zeppelin as Germany's premier airship creator. After World War I ended, Eckener improved airship design, hoping to surpass his successful *Graf Zeppelin* by developing the largest, fastest airship ever: an airship that would be better suited for long-distance flight and be comfortable, safe, and profitable as well.

German technology is renowned for safety

Eckener was about to begin the new airship's construction at the end of 1934. By this time, every major nation of the world that had an airship program had either abandoned it or planned to abandon it. Each of these countries—except Germany—had experienced fatal crashes. But German airships had no serious accidents and no passenger fatalities; they enjoyed a spotless record in commercial airship flying.

The *Hindenburg* was named after Field Marshal Paul von Hindenburg, a respected German president and war hero. This latest airship measured 803 feet, 29 feet longer than the *Graf Zeppelin*, but it was far wider than its predecessor—135 feet in diameter. It could hold nearly twice the gas as the *Graf Zeppelin*, giving it greater lifting power and making it more efficient and less prone to bending. It was equipped with the latest Daimler-Benz diesel engines: lightweight, yet—at 1,000 horsepower—sufficient to drive the ship to 85 miles per hour.

Airship delivers speed and luxury

The new *Hindenburg* made the westbound trip in 65 hours, and the eastbound route, from Lakehurst, New Jersey, to Frankfurt, Germany, took just 52 hours—far faster than any ocean liner. Yet the *Hindenburg* was in fact a flying luxury liner. For the one-way fare of $400, a passenger could have a comfortable private cabin with hot and cold running water. The 50-foot long dining room joined the 50-foot promenade deck with wide, sloping picture windows. The lounge on the starboard (right) side

The *Hindenburg* crashes to the ground in Lakehurst, New Jersey.

was furnished with tables and comfortable upholstered chairs and measured 34 feet long. There was also a writing room and another promenade deck. The passengers enjoyed spacious toilets, washrooms, and showers.

Designed for helium, but filled with hydrogen

The *Hindenburg* was designed to be lifted by helium, but the United States refused to sell its scarce helium to the Zeppelin airship company. So the *Hindenburg* had to be filled with hydrogen—7 million cubic feet of the extremely flammable gas. Fire on an airship was of intense concern—the crew impounded all matches from passengers as they boarded. (Smoking was allowed during flight, but only in a room specially sealed and pressurized.)

Hitler allows no test flights—rushes the *Hindenburg* into service

The United States suspected that Germany's new leader, Adolf Hitler, might one day have military plans for helium-filled airships. In fact, the

Hindenburg's maiden flight, on March 4, 1936, was actually a propaganda mission over Germany—the airship spent four days and three nights broadcasting speeches and dropping leaflets in favor of Hitler's policies. The propaganda flights prevented the airship's normal test trials, which distressed designer Hugo Eckener.

Hitler knew that the very popular Eckener was firmly opposed to Nazi principles, so he "promoted" the airship designer to chairman of the board, a position with no real power. Then Hitler's government became half owner of Eckener's company.

Hindenburg flies well—even in bad weather

The United States granted permission in February 1936 for the *Hindenburg* to make ten round trips to Lakehurst, New Jersey. The *Hindenburg* successfully completed all of these flights on schedule without difficulties. The first flight to Lakehurst took 2½ days, and its return took only 2 days, 1 hour, 14 minutes. The *Hindenburg* also made six trips to Rio de Janeiro, Brazil. Its 1936 records show that it carried more than 1,500 transatlantic passengers and 20 tons of mail and freight. Travelers endorsed the airship enthusiastically, and American financiers were showing interest. The ship proved to be amazingly stable during its transatlantic crossings, even during gale-force winds. (It actually weathered a hurricane without the passengers being aware of it.) The plush accommodations, the sumptuous meals served on china, and the lack of seasickness indicated a bright future for airship travel. A German-American international company made plans to put four Zeppelins into weekly service across the Atlantic.

Ill-fated Last Flying Season Begins

The *Hindenburg*'s 1937 flying season began in March with a flight to Rio. The *Hindenburg*'s U.S. flying season was set for May through November, with the first of the 18 round trips between Frankfurt and Lakehurst scheduled for a May 3, 1937, departure. Previous bookings had been full, but the passenger list for this flight totaled only 36, most of whom were German officers and trainees. The Zeppelin company planned to launch the *Hindenburg*'s sister ship, the LZ-130, for its first flight in October 1937.

Max Pruss, former captain of the company's *Graf Zeppelin,* was in command as the *Hindenburg* cast off at 8:15 P.M. from the new Rhein-Main World Airport near Frankfurt. The airship headed north and cruised at nearly 90 miles per hour. Off the coast of Ireland it encountered a strong headwind.

The ship avoided flying over France or England for political reasons, flew over the Netherlands to get to the North Atlantic, then down the English Channel. Its arrival time at Lakehurst was to be May 6 at 6:00 A.M., but headwinds reduced its speed to 60 miles per hour. Further delays occurred as the weather deteriorated around Newfoundland, Canada, reducing the ship's speed to 37 miles per hour as it reached the New England coast. Its arrival time was now estimated to be around 6:00 P.M. on May 6, a full 12 hours behind schedule.

Hindenburg ordered to come in

It arrived over New York City at 3:00 P.M., but an approaching cold front forced the *Hindenburg* to ride out more local bad weather by staying at sea. After heavy showers and a thunderstorm, conditions improved, and Lakehurst gave the order for the *Hindenburg* to come in. It finally came up the field at Lakehurst from the southwest just after 7:00 P.M.

At 7:08 P.M. the *Hindenburg* emerged impressively from the clouds at 650 feet, roaring over Lakehurst at full speed. It made a sweeping turn to the left to approach the mooring from the west. As the airship returned over the field at 7:10, shifting winds convinced Pruss to decide on a northerly approach, for which he had to make another much sharper turn to the right. The ceiling was between 2,000 and 3,000 feet, a light rain was falling, and lightning was noticed in the distant south and southwest. Pruss valved off some hydrogen, lowering the ship a bit, and he dropped some water ballast. By maneuvering his four engines, he brought the huge ship to a complete, dead-level stop in midair at 7:20 P.M.

At 7:21, with the ship at about 200 feet altitude and about 700 feet from the mooring mast, crewmen dropped the starboard (right) ropes to the ground crew, then they dropped the port (left) lines. The landing was proceeding normally as the ground crew joined the ropes to the lines on the ground.

Skin flutters! Flames emerge!

The ship was hovering between 135 feet and 150 feet when its outer cover began to flutter and its skin seemed to be rippling. About 15 seconds later—at 7:25 P.M.—a small tongue of flame emerged where the skin had been fluttering and a reddish glow was noticeable from the ground. In the control car the crew felt a shudder in the ship's frame. At the same

The skeleton of the *Hindenburg* after it crashed; the disaster ended airship travel forever.

time a member of the crew in another part of the ship heard a muffled pop, like a gas burner igniting on a stove.

Completely destroyed—in just 32 seconds

Seconds later a burst of exploding, flaming hydrogen blasted out of the top, just ahead of the upper fin. Within a few seconds, nearly the entire stern was engulfed in flames and began to drop. Inside the control car Pruss made the split-second decision to let the ship's tail section crash to the ground, knowing this was the only hope of anyone getting out alive. As the great ship's tail dropped to the ground, its nose pointed skyward. Flames shot up through promenades, which acted like chimneys, and spewed out the nose "as from a blowtorch." Now the entire ship was burning, and its framework quickly collapsed. From the time the first flame was noticed until the entire ship lay smoldering on the ground, just 32 seconds elapsed.

The fire killed 13 passengers, 22 crewmen, and 1 civilian ground handler. Of the 97 people on board, 62 amazingly survived. Some men in the tail walked out, virtually untouched, as the flames in that section went upward. A cabin boy was saved when he was doused by ballast water. An acrobat hung from a window as the ship fell and let go at a safe height. Others were saved by the heroic actions of the ground crew. Captain Pruss refused to leave and went back several times to help survivors get out. The *Hindenburg* burned for three hours.

Impact

The reasons for the crash of the *Hindenburg* remain mysterious and controversial. Of all the explanations over the years about what caused it suddenly to catch fire, two major theories persist: ignition of hydrogen by sabotage, or ignition by some natural source of electricity. Neither the German nor the American investigation could produce any evidence of sabotage, although some argue that any admission of sabotage would have created an embarrassing international incident. A book published in 1962 named the alleged saboteur aboard the airship, but to this day no conclusive proof exists that the airship was sabotaged.

Examiners consider several theories

Several natural explanations were offered, and it was one of these that the German and American investigating teams chose to endorse. Both teams of experts concluded that the airship's hydrogen was ignited probably by some type of atmospheric electrical discharge. The Americans argued that St. Elmo's Fire—a discharge of electricity that sometimes occurs as an eerie bluish glow on the prominent parts of a ship or aircraft in stormy weather—ignited the hydrogen. The Germans said that the ropes dropped to the ground became wet, resulting in the airship "becoming a piece of ground elevated into the atmosphere." This equalization of the static charges between the ship and the ground meant that the *Hindenburg* would itself discharge electricity into the atmosphere, a phenomenon known as "brush discharge."

Hindenburg explosion is officially ruled "accidental"

Both natural-cause arguments assume that free hydrogen existed somewhere around the ship and was available to be ignited. The fluttering cover is said to prove that hydrogen was escaping. Bolstering this observation is the ship's tail-heaviness, witnessed by most just before the explosion, which may have indicated possible hydrogen loss. The official judgment is that the *Hindenburg* was destroyed accidentally by unusual but natural causes. Designer Hugo Eckener agreed with this verdict, but Captain Pruss argued for sabotage. The debate continues even today, with some recent information adding weight on the side of the natural causes: the *Hindenburg*'s surface had been painted with a different type of aircraft dope, or surface preparation material, which ultimately helped create the deadly spark.

How the Crash Influenced History

At the time of the *Hindenburg* crash, the German airship *Graf Zeppelin* was flying back from Rio. On its arrival in Germany, the ship was grounded until the cause of the *Hindenburg* crash could be determined. No Zeppelin ever made another flight. The explosion of the *Hindenburg* marked the end of hydrogen as a lifting medium and the end of airship travel.

Despite the fact that for more than a quarter century commercial Zeppelins had carried 50,000 passengers without a fatality, airship travel was suddenly abandoned. This result must surely be due to the *Hindenburg* being the most thoroughly documented crash of its time. The sights and sounds of the *Hindenburg* fire were captured on film and played over and over again in movie newsreels. On radio, the heartbreaking, raw emotion of newsman Herb Morrison—"Oh, the humanity and all the passengers!"—conveyed the horror and pity he felt as he watched helplessly.

The spectacular burning of the *Hindenburg* affected public opinion far more than its fatality count. Despite the fact that nearly two-thirds of the people on board survived, its name became linked forever with tragedy and sudden, terrifying technological disaster. Nothing could ever be done to erase the disastrous image of airship travel. The *Hindenburg* disaster ended airship travel forever, but two years later commercial transatlantic service was available again—by airplane.

Where to Learn More

Dick, Harold G. *The Golden Age of the Great Passenger Airships: "Graf Zeppelin" and "Hindenburg."* Washington, DC: Smithsonian Institution Press, 1985.

Hoehling, Adolph A. *Who Destroyed the "Hindenburg"?* Boston, MA: Little Brown, 1962.

"Oh, the Humanity! *Hindenburg* Disaster." *Time* (May 17, 1937): 35ff.

Payne, Lee. *Lighter than Air: An Illustrated History of the Airship.* New York: Orion Books, 1991, pp. 218–29.

"Sky Horror: *Hindenburg* Crash." *Literary Digest* (May 15, 1937): 10–12.

Vaeth, J. Gordon. "What Happened to the *Hindenburg*?" *Weatherwise* (December 1990): 315–22.

BOAC Comets explode

Mediterranean Sea, near the island of Elba
January 10, 1954

Mediterranean Sea, near the island of Stromboli
April 8, 1954

Two new jetliners explode in midair due to cracks that developed around small windows in the upper fuselage. The cycles of pressurization and depressurization that are normal to jet travel are found to cause fatal metal fatigue.

Background

On January 10, 1954, a Comet jetliner owned by British Overseas Airways Corporation (BOAC) was flying out of Rome on its way to London. At 10:50 A.M. it radioed the Rome air traffic control tower that it was breaking through the overcast at 27,000 feet and was climbing to its assigned cruising altitude of 36,000 feet. Less than two minutes later, fishermen off the island of Elba in the Mediterranean saw flaming wreckage fall from the clouds into the sea.

Thirty-five lives are lost

Registered as G-ALYP (when talking with the tower, the pilot pronounces the call letters "George Yoke Peter"), the aircraft radioed a routine message to another BOAC flight regarding the height of the cloud cover. That report was interrupted in midsentence. Suddenly 29 passengers and 6 crew members vanished.

British are first to offer commercial jet service

The age of jet transportation was only two years old—begun when BOAC instituted scheduled jet service. BOAC flew the same aircraft to South Africa. The original Comet accommodated only 36 passengers (40 on a model designed with extra fuel tanks for greater distances), but it revolutionized air travel with its speed and ability to fly above virtually any kind of weather. England was first to achieve this technological triumph—Russia was two years away from launching its TU-104 (actually a converted bomber), and America's Boeing 707 and Douglas DC-8 were

Both Italian and British authorities attempted to determine the cause of the "mysterious" crash. Thirty-five persons met their deaths when the jet went down near Elba on January 10, 1954.

still on their drawing boards. De Havilland, the Comet's builder, had firm airline orders for 50 Comets, and negotiations for another 100 were well underway. Then tragedy struck.

Takeoff accidents blamed on pilot error

Comet jet service enjoyed enormous success for the first two years, but there were two takeoff accidents. One was not fatal, but the other resulted in the death of everyone on board. In each incident the pilot attempted to rotate (lift) the nose before attaining adequate speed. In a third accident, a BOAC Comet flew into a violent thunderstorm and apparently disintegrated in the storm's lightning and turbulence. Yet neither fatal crash seemed to raise any fears—aviation had a long record of failures due to pilot error and storms. Flaws in the Comet's design were never thought to be factors in any of the accidents.

However, chief designer Ronald Bishop and his engineering staff were well aware that the new jetliner operated at unprecedented speeds and altitudes. Cabin pressurization, which was known to stress aircraft structure, dated back to the Boeing Stratoliner of 1938. The thin air of the upper atmosphere was compressed into heavier, more breathable air as it entered the cabin. But the pressurization and depressurization cycles of piston-engine planes differed markedly from those of jets, which climbed and flew twice as fast and routinely cruised 10,000 feet higher.

Engineers exceed structural standards

The aircraft's central compartment, the fuselage, houses the crew, passengers, and cargo. The fuselage of the propeller-driven Constellation, the DC-6, and other transport aircraft was designed to safely hold a maximum pressure of about 4 psi (pounds per square inch). This means that at the altitude of 25,000 feet, the air pressure in the cabin actually equals an altitude of less than 5,000 feet. The Comet required 8 psi to simulate the breathable air of the lower altitude, but British civil aviation authorities insisted that the Comet's cabin structure cope with more than 16 psi. The de Havilland engineers met that standard and more: Bishop and his chief structural engineer, Robert Harper, designed the Comet fuselage to withstand 20 psi.

Jetliner's skin made of thin-gauge aluminum

To minimize aircraft weight, the engineers chose a relatively thin-gauge aluminum. The skin was only 0.028 (28/1,000) inch thick, not much thicker than a postcard. Window frames, which were squared off, proved to be a controversial—and critical—design decision. The U.S. Civil Aeronautics Administration (CAA) questioned whether these windows were durable. This agency—which later became the Federal Aviation Administration (FAA)—was supposed to determine whether American-operated Comets were airworthy, and it suggested that oval-shaped frames would distribute pressurization stresses more equally. Nevertheless, BOAC engi-

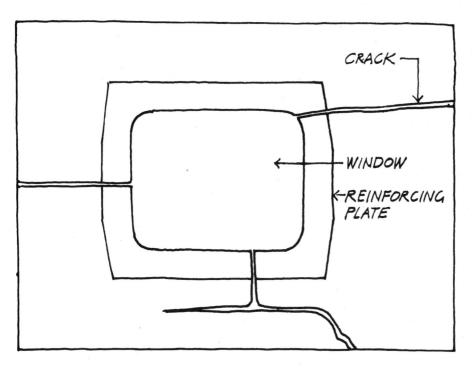

CRACK

WINDOW

←REINFORCING PLATE

Metal fatigue cracks around a small window in the upper fuselage developed during normal pressurization and depressurization cycles. These cracks caused fatal cabin depressurization.

neers maintained that the Comet's windows had been tested at up to 100 psi without any signs of fatigue.

After G-ALYP crashed over the Mediterranean island of Elba, BOAC voluntarily grounded all seven aircraft in its Comet fleet for inspection. This precautionary measure gave investigators a chance to do their work. The scenarios they considered included:

- Sabotage
- Explosion caused by a ruptured turbine blade penetrating a fuel tank
- Structural failure from clear air turbulence
- In-flight engine fire that either ignited fuel or weakened the structure to the point of failure
- Explosion of hydrogen from a leaking battery
- Explosion of fuel vapor in an empty tank.

To minimize aircraft weight, the engineers chose a relatively thin-gauge aluminum. The skin was only 0.028 inch thick, not much thicker than a postcard. Window frames, which were squared off, proved to be a controversial— and critical— design decision. The Civil Aeronautics Administration suggested that oval-shaped frames would distribute pressurization stresses more equally.

Could so new an aircraft suffer metal fatigue?

One technical expert at BOAC wondered if metal fatigue might have caused explosive decompression. However, company officials considered the suggestion impossible. Comet G-ALYP had logged fewer than 4,000 hours in the air, which is equivalent to driving an automobile around for a few months. The designers at de Havilland assured BOAC that, structurally, Comets could endure the stress of repeated pressurization and depressurization—even 10,000 times.

Salvage crews seek Comet's wreckage

Investigators were stymied in their search for the cause of the crash because they were short on solid clues. Royal Navy salvage crews groped for wreckage buried 500 feet deep and ranging over 100 square miles of Mediterranean sea floor. Back on dry land, however, a special investigating committee recommended some 50 modifications to the Comet. Considered to be preventive measures, they were based on possible—but unproven—midair explosion theories. Among the changes implemented were these:

- Strengthening fuel lines
- Installing armor-plated shields between the engines and fuel tanks
- Replacing original smoke and fire detectors with improved equipment.

Another Comet claims 21 lives

By late March 1954, BOAC's improved Comet fleet was flying again, even though salvage crews were still looking for hard evidence. Despite the improvements made to the jetliner's design, though, a second Comet aircraft exploded on the night of April 8, 1954, and 21 people died. This aircraft, with call letters G-ALYY ("George Yoke Yoke"), was flying from Rome to Cairo. As it climbed to its assigned altitude of 35,500 feet, radio contact suddenly ended. The next morning, 5 bodies and 2 aircraft seats were recovered from the sea near the island of Stromboli. For the second time in less than four months all Comets were grounded. British prime minister Winston Churchill ordered the Royal Aircraft Establishment (RAE), England's prestigious organization of aeronautical science headed by Sir Arnold Hall, to investigate both accidents. What the RAE turned up about why the world's first jetliners were exploding also determined the fate of all future jet travel.

Details of the Disaster

The scientists of the RAE owe their eventual discovery to the salvage operations of the Royal Navy. Yoke Yoke's wreckage had plunged into water 3,500 feet deep—chances for finding any debris were remote. But less than a month after Yoke Yoke disappeared, the navy recovered about two-thirds of Yoke Peter's wreckage. The remnants were put on a ship bound for Farnborough, the main RAE research facility. RAE chief Arnold Hall could not tolerate the slow delivery schedules, so, without waiting for official permission, he commandeered a huge U.S. Air Force cargo plane to deliver all the salvaged scrap.

England's aeronautical chief suspects metal fatigue

Almost from the start Hall suspected metal fatigue—and so did Dr. Peter Walker, head of RAE's Structural Department. The investigation detoured temporarily down one blind alley when they discovered that one of Yoke Peter's engine turbine blades was missing. This finding made the investigators lean toward the theory that a severed blade had punctured a fuel tank. But this possibility was discarded after the turbine casing was found intact and it was determined that the blade had been torn off by the impact of the crash.

Autopsies support the RAE's suspicions

Then came the final reports from the autopsies performed in Italy on the 15 bodies recovered from Yoke Peter and the 5 found near the site of the Yoke Yoke crash. The information suggested to the RAE that all had been victims of "violent movement and explosive decompression." The Italian pathologist concluded that the traumatic ruptures of the hearts and lungs of the victims occurred before impact into the sea, a finding that supported the RAE investigators' suspicions: the pressure in the cabin failed prior to the explosion.

Hall ordered construction of a tank large enough to hold a Comet fuselage. One of the grounded jetliners was placed inside, with its wings protruding from a hole on each side of the tank. Hall believed that the original pressurization tests conducted on the Comet prototype were inadequate. True, the Comets were subjected to repeated pressurization and depressurization cycles to determine what effect they would have on the structural life of the aircraft, but the testing had not used hydraulic jacks, which would simulate the motion of an aircraft in flight.

RAE uses hydraulic jacks in new test

The RAE used the jacks to flex the wings of the submerged Comet. The fuselage was filled with water (instead of air, because water would cushion an explosive decompression), and the pressure was raised to slightly over 8 psi (pounds per square inch), which is the Comet's normal pressurization at 35,000 feet. Pressure of 8 psi was maintained for three minutes while wings were moved up and down, after which pressure was reduced to 0 psi, then increased back up to 8 psi for another three minutes. This cycle, corresponding to the pressure variations of a three-hour flight, was repeated around the clock until the airframe had accumulated the equivalent of 9,000 flying hours (or 3,000 flights). The test schedule aged the airframe 40 times faster than would have been possible in normal airline service.

Fuselage fails new test

In late June 1954, just as the aircraft reached the 9,000-hour mark, the fuselage lost pressure and cracked. After the water was drained from the tank, investigators discovered a crack 8 feet long along the top of the fuselage that extended from a fracture in one corner of a small escape-hatch window and bisected a cabin window frame. Metallurgic inspection of the hatch window revealed discoloration and crystallization, the telltale signs of metal fatigue.

In mid-August salvage crews recovered a section of Yoke Peter's fuselage roof and flew it to Farnborough. The roof bore the same damage that occurred during the testing: In the corner of a navigation window atop the fuselage was an ominous crack that had grown into a wide split. With this final clue, RAE investigators concluded that the accidents were primarily "caused by structural failure of the pressure cabin, brought about by fatigue."

RAE claims several factors caused crash

As with most air disaster investigations, the solution to the Comet mystery produced more than one culprit and no single cause. The first factor leading to the Comet crashes was an inadequate test program, which did not accurately gauge the long-range effects of continued pressurization and depressurization cycles on airframe integrity. Square-shaped window frames and the absence of a metal-fatigue prevention program, which would prevent fatigue cracks from spreading, compounded the disaster. The third fatal factor was the thin fuselage skin—so thin that one U.S. airline president, on an early Comet demonstration ride, swore he could see the sides of the cabin walls "moving in and out like an accordion."

Impact

Three versions of the Comet were built—MK-1, MK-2, and MK-3. The two Mediterranean crashes involved only the MK-1 version. MK-2 Comets, larger and more powerful and carrying up to 44 passengers, were changed extensively after the RAE investigation. A number of European airlines as well as the Royal Canadian Air Force used MK-2s. But de Havilland built only one Comet 3—the lingering stigma of the 1954 disasters and the development of new, larger, American-built jetliners proved fatal to the Comet's market life.

Fly jets "by the book"

In the wake of the Comet investigations, people assumed that the designers of the newer jets, the Boeing 707 and the Douglas DC-8, learned valuable lessons that would result in new safeguards against metal fatigue and explosive decompression. In truth the industry learned more from the Comet's pair of takeoff accidents than the disasters at Elba and Stromboli. They found out that it was absolutely necessary to fly jet transports "by the book," as well as to test new aircraft design. Even before the Comet tragedies, both Boeing and Douglas were already designing jet fuselages with:

- Thicker skin
- Triple-strength, rounded windows to distribute pressurization stresses equally around the frames
- Metal bracing to provide additional structural strength, similar to reinforcing a wooden barrel with iron staves

and the most important insurance against catastrophic decompression:

- Small metal tabs, or "stoppers," strategically placed throughout the fuselage

If a fatigue crack should develop, its path would be blocked before explosive decompression could occur. All rebuilt Comet 2s and the single MK-3 also incorporated these structural reinforcements.

Boeing demonstrates the effectiveness of "stoppers" in 707s

Boeing dramatically demonstrated the efficiency of this preventive system before the 707 ever flew. Engineers deliberately weakened a 707 test fuselage with saw cuts up to 22 inches long, then pressurized the cabin and dropped 5 huge, stainless steel blades, which slashed into the top of the fuselage. Small puffs of air escaped from the wounds, but no catastrophic explosion occurred.

As an introduction to its successful preventive system, Boeing dramatized to airline officials what happens to a pressurized cabin without stoppers. They dropped two steel blades onto a fuselage without the metal tabs. Filmed at high speed and replayed in slow motion, the motion picture showed the metal fuselage skin begin to curl outward at the points of penetration, then curl faster until the entire cabin split open and spit out its contents—seats, dummy bodies, and even the cabin floor. It vividly portrayed what had happened to the Comets.

Where to Learn More

Anderton, David. "How the Comet Mystery Was Solved, Part II: Giant 'Jigsaw Puzzle' Gives Final Clue." *Aviation Week* (February 14, 1955): 26–39.

———. "RAE Engineers Solve Comet Mystery." *Aviation Week* (February 7, 1955): 28–42.

"Comet Crashes: Dakar Accident." *Aviation Week* (July 27,1953): 16.

"Comet Crash off Elba Attributed by British Inquiry to Metal Stress." *New York Times* (October 20, 1954): 14.

"Death of the Comet." *Time* (April 19, 1954): 31–32.

Dempster, Derek D. *The Tale of the Comet.* New York: David McKay, 1958.

Hull, Seabrook. "Fatigue Blamed in Comet Crashes." *Aviation Week* (October 25, 1954): 17–18.

McKitterick, N. "Comet Crash." *Aviation Week* (May 11, 1953): 17.

———. "Comet Future?" *Aviation Week* (January 19, 1954): 16–17.

"What's Wrong with the Comet?" *Business Week* (April 17, 1954): 27–28.

"Why the Jets' Skin Ruptured." *Business Week* (November 6, 1954): 121–124.

Lockheed Electras crash

Over Buffalo, Texas
September 29, 1959

Near Tell City, Indiana
March 17, 1960

Background

On September 29, 1959, flying in clear, calm weather over Buffalo, Texas, a new Braniff Lockheed Electra lost a wing and crashed. All 33 people on board—28 passengers and a crew of 5—lost their lives in the crash. After investigating fruitlessly for five and a half months, stumped examiners were ready to write off the Braniff accident as "crashed for reasons unknown."

Then, on March 17, 1960, a Lockheed Electra owned by Northwest also lost a wing in flight and crashed near Tell City, Indiana. All 63 on board were killed. The Civil Aeronautics Board Bureau of Safety (now the National Transportation Safety Board) now faced two unexplainable fatal crashes of a new aircraft that was, they thought, thoroughly tested.

Ninety-six people perish when two brand-new turboprops suffer wing failure due to a type of flutter—whirl mode—previously regarded as harmless.

Lockheed Electra propjets pass rigorous pre-service testing

The Lockheed Electra L-188 was the first U.S.-designed turboprop transport—more commonly called a "propjet"—to enter airline service. The Electra had four Allison jet engines that were hitched to huge, four-bladed propellers. This gave the Electra a cruising speed almost equal to pure jets. Pilots loved its enormous reserve power and responsiveness.

Although the world's airlines were falling in love with big jets, there was still a viable market for a fast, versatile propjet. Lockheed, located in Burbank, California, planned to meet the demand with its Electra L-188, and it took orders for nearly 180 Electras from 16 U.S. and foreign air carriers.

The Braniff Lockheed Electra exploded in midair, leaving this tail section the largest piece of debris. Eyewitnesses reported seeing a giant fireball at around 15,000 feet, the flight's assigned cruising altitude.

Development costs for the L-188 amounted to some $50 million, with four years of engineering research preceding the Electra's first flight. Passing the rigorous testing and certification process made the subsequent Braniff and Northwest accidents difficult to understand.

Braniff propjet crash stumps investigators

At the Braniff site the investigators from the CAB (Civil Aeronautics Board) were stumped. The flight engineer's log sheet, recovered from the cockpit wreckage, did not supply much data: only the aircraft's speed and altitude seven minutes before some unknown menace struck. They knew structural failure was a factor—they found the left wing and both its engines one and a half miles from the potato patch where most of the rest of the wreckage lay. The debris was strewn along a path nearly 14,000 feet long, and from the way the wreckage was distributed, investigators judged that the left wing failed first, followed by general disintegration of the entire aircraft. Eyewitnesses reported seeing a giant fireball at around 15,000 feet, the flight's assigned cruising altitude.

The search turned up few clues that were of any use. Investigators considered—and one by one eliminated—every possible explanation for the Braniff crash. Among the crash theories were these:

- An unchecked in-flight fire

- An out-of-control autopilot

- An excessively violent maneuver to avoid another plane

- An accidentally fired missile

- A dive-inducing runaway propeller, causing a startled pilot to pull back on the controls too abruptly

- The remote possibility that an overheated tire or wheel brake somehow ruptured a fuel line.

Before propjet crashed, "every coon dog for miles around started howling"

Examiners couldn't find conclusive evidence to support any of the theories, but one peculiar circumstance nagged at everyone. Farmers reported that just before they saw the fireball in the sky, "every coon dog for miles around started howling." Other witnesses told of hearing a variety of unusual noises, ranging from thunder to sonic booms, from a "whooshing, screaming noise" to "the clapping of two big boards together." The CAB team went to work recording every known source of intense aeronautical noise—jet aircraft, sonic booms, runaway props, as well as Electras in normal flight and in dives and climbs. The tapes, which included intentionally random noises not associated with airplanes, were played back to witnesses who were asked to select the sound or sounds closest to what they heard. Most people picked two:

- A runaway propeller whirling at supersonic speeds

- And a jet airplane.

Northwest propjet crash in March helps solve puzzle

The significance of the eyewitnesses' choices wasn't grasped immediately. In fact, the Civil Aeronautics Board was about to inform Braniff that it could release the remains of the aircraft to its insurance underwriters when another Electra crashed.

The Northwest Electra wing failure caused the propjet to go down near Tell City, Indiana, and killed all 63 people on board. The two accidents were not identical. The Northwest propjet was flying higher, at 18,000 feet in an area of reported clear-air turbulence, and it was the right wing that failed first. But in the two piles of wreckage the CAB found certain similarities that added up to one conclusion: both planes were victims of destructive flutter. Flutter is the oscillation, or vibration, of an aircraft part caused by the airstream. This is similar to what happens when a lightweight object is held out of an open window while a car is moving.

Details of the Crashes

The March disaster to the Northwest aircraft produced the clearest evidence. Investigators found numerous indications of damage on the right wing structure between the fuselage and the inboard nacelle (engine housing). Rapid reversals of load forces had progressively damaged the wing—in other words, the wing experienced massive, uncontrolled flutter severe enough to cause it to separate from the fuselage. In fact, the metal damage indicated that the wing folded rearward—not upward. Upward folding has been documented in a high up-gust or in a positive maneuver. The Northwest propjet had saw-toothed diagonal fractures in a key area of the wing's structure, which indicated that powerful oscillation loads—back and forth movement—had occurred repeatedly. The evidence proved catastrophic flutter.

Scars in the metal prove turboprops were destroyed by flutter

Some vibrating source produced abrasions in the metal on the outlet of the sugar scoop, which is an air inlet. The mounts that held the huge Allison engine in the outboard nacelles had "bottomed," or shaken, up to 60 times at a rate of 2.5 times a second for up to 20 seconds. Marks on the engine torque housing indicated that there had been repeated cycling at the bolt attachments. A vane located near the engine exhaust was scarred with elliptical scratches, which is telltale evidence that the engine casing had sustained violent motion.

Examiners went back to the Electra flown by Braniff, the first of the two Electras to fail. Evidence of destructive flutter was not as clear-cut, but there were significant duplications. Flutter damage was greatest in

The Northwest crash in Tell City produced the clearest evidence that severe flutter of the right wing tragically caused wing separation.

the outboard engine on the failed wing of each aircraft. Both showed severe engine mount oscillation. Both showed marks of repeated, massive vibration in the outboard nacelle structure. Yet blaming both crashes on flutter left a major question unanswered: how could airline personnel have allowed flutter to progress to the point of structural failure?

All aircraft experience flutter

Flutter is caused by air turbulence over fast-moving airplane surfaces: it is present in every type of aircraft. Design engineers aim to keep it in check—to damp it—because continuous, uncontrolled fluttering can produce fatigue even in new metal.

Precertification tests showed the Electra to have excellent flutter-damping characteristics at all speeds. If an external driving force such as turbulence were to "excite" the wing, the resulting vibration could be damped by a slight reduction in speed. Furthermore, the Electra's wing, like that on all airplanes, was designed to flex and recoil, thus absorbing energy from flutter oscillation.

After the CAB determined that flutter was involved in both accidents, it faced another riddle: why did different turbulence conditions result in the same type of failure? Severe clear air turbulence was a factor in the Indiana crash in March, but there was absolutely no turbulence present in the September skies over Texas. Investigators begin searching for a flutter-inducing source other than turbulence.

Test flyers deliberately invite trouble

Wreckage from the Buffalo and Tell City crashes was shipped to Burbank for renewed examination. Meanwhile Lockheed pilots crammed an Electra with instruments and, deliberately courting destructive flutter, took her through a series of nearly 70 hair-raising test flights, most of them into areas of known severe turbulence. Boeing provided a team of flutter experts to help Lockheed interpret the data from the test flights. Douglas lent its California competitor a new flutter-inducing device—vanes mounted on the wings. The angles of the vanes were controlled from the cockpit. These missions totaled some 1,500 hours of dangerous flying and produced the first evidence of trouble in the outboard nacelles: severe and extreme turbulence exerted a bending force on the outer wing section, from the outboard nacelles to the tips, that was 10 times greater than on the rest of the wing.

Lockheed and the CAB added this revelation to the signs of flutter damage found in the two wrecks, and they began searching for the type of flutter that could destroy a sturdy wing. There are more than 100 kinds of flutter, called modes. The experts went back to their tape recordings and zeroed in on the sound of an overspeeding propeller. They tried every flutter mode capable of producing such potentially destructive speed and found exactly one in which the propeller tips approached sonic velocity but showed no increase in revolutions per minute or air speed: whirl mode.

Whirl mode is a vibration natural to any piece of rotating machinery, including airplane propellers. Until the Electra crashes, however, whirl mode was not considered even remotely hazardous. Damping checked all flutter modes in piston airplanes, because if flutter continued, the prop would tear itself off. But why had damping failed in the L-188?

Examiners found the answer in the nacelles, or housing structures. In both aircraft:

- Support struts had been bent and severed
- Engine mounts had been subjected to abnormal loads in various directions
- Fractures in the propeller reduction box (which keeps the prop from revolving as fast as the turbines) displayed signs of vicious multidirectional loads, and
- There were curved scratches on certain engine parts

This proved that there were huge cyclical motions—not just of a wobbling propeller—but of virtually the entire engine.

Wings fail in wind-tunnel testing

Investigators then performed wind tunnel experiments on a scale model of an Electra, which reproduced the same effects that occurred over Buffalo and Tell City. The experts deliberately weakened the struts and braces, which would normally hold the "engine package" firmly in place. "Flying" the model at speeds equal to or greater than those of the Braniff and Northwest planes, the technicians created a sudden jolt (such as simulated turbulence). The jolt excited the outboard nacelles, or engine housings. Whirl mode started, fed itself on its own increasing energy, and eventually devoured the wing.

A propjet "engine package" is, in effect, a huge gyroscope, whose turbines are spinning at nearly 14,000 rpm (revolutions per minute) while the prop is spinning at more than 1,200 rpm. The jolt that excited the outboard nacelles was like a giant finger reaching out to touch this smoothly whirling mass, which then broke stride and wobbled into a flutter mode. The flutter gained new strength from every unending cycle of violent motion until the wing itself surrendered.

Impact

The Northwest crash in March over Tell City prompted the Federal Aviation Administration (FAA) to order all U.S.-operated Electras to fly at lower speeds until the CAB could establish the cause of the two accidents. After the cause was determined, speed restrictions remained in effect until Lockheed could prove that its $25 million modification program was effective.

The fix called for numerous structural changes: approximately 1,400 pounds of strategically placed reinforcements went into the wings as well as the nacelles. In the most daring test flight of all, Lockheed pilots flew an Electra into an area of severe turbulence and deliberately induced whirl mode. Douglas vanes verified that the structural improvements indeed damped the "death mode." Speed restrictions on modified Electras were lifted.

Though unchecked whirl mode in a weakened nacelle caused the wings to fail, what weakened the Braniff and Northwest outboard nacelles in the first place? The original weakening was never conclusively established. A previous hard landing was suspected in the case of the Northwest Electra, but the source of earlier damage on the Braniff plane remains unknown.

Pre-impact fatigue breaks are found in both aircraft engines

Another question remains unanswered: was the Electra's vulnerable area the engine or the wing? Allison, the manufacturer of the jet engines, claimed that whirl mode should not have fatally affected the structural integrity of the wing, unless wing and nacelle strength were marginal to begin with. Lockheed argued that whirl mode could not be transmitted to a wing solely through a weakened nacelle, that whirl mode was possible only if a failure occurred first within the engine itself. In the outboard engines of both aircraft, investigators found that fatigue breaks existed in the gear boxes and power sections prior to the propjets crashing.

The dispute over the Electra's area of vulnerability may never be resolved.

Where to Learn More

"Braniff Electra Crash Investigated by CAB." *Aviation Week and Space Technology* (October 5, 1959): 38–39.

Bulban, Erwin J. "Braniff Electra Disintegrated in Flight, CAB Investigator Says." *Aviation Week and Space Technology* (October 26, 1959): 45.

"CAB Digs Deeper into Electra Troubles." *Product Engineer* (August 29, 1960).

"Electra Accidents." *Engineer* (July 1, 1960): 45–46.

"Lockheed Completes Electra Test: Announces Probable Crash Cause." *Aviation Week and Space Technology* (May 16, 1960): 40.

Mooney, Richard E. "Parley Will Sift Electra Crashes." *New York Times* (March 22, 1960): 74.

"Thirty-three Feared Killed in Airliner Crash." *New York Times* (September 30, 1959): 1.

United Airlines DC-8 and TWA Constellation collide

Over Staten Island, New York
December 16, 1960

Background

Midmorning on December 16, 1960, a bleak winter sky hung over the metropolitan New York area. A United Airlines DC-8 jetliner, carrying 77 passengers and 7 crew members, was flying in from Chicago. Nearing Staten Island at an altitude of 5,000 feet, the jet was soon to land at Idlewild Airport. Meanwhile, a prop-driven Constellation (called a "Connie"), owned by Trans World Airlines (TWA), was also approaching Staten Island. Carrying 39 passengers and 5 crew members, the TWA craft was heading for LaGuardia Airport after a routine flight from Columbus, Ohio. The air traffic controller watched the blip that represented the Connie move steadily on the radar scope in LaGuardia approach control. Then the controller saw another blip at 5,000 feet—the same altitude as the jet.

A midair collision over Staten Island between a United Airlines DC-8 and a prop-driven TWA Constellation exposes glaring weaknesses in the nation's "completely modernized" air traffic control system.

Radar blips merge on the control screen . . .

"Unidentified target approaching you . . . six miles . . . jet traffic," he advised the TWA flight.

"Roger, acknowledged."

The controller kept his eyes on the two blips, frowning as he watched their progress. He paged TWA again.

"Identified object . . . three miles . . . two o'clock."

"Roger. Acknowledged."

Before the controller's horrified eyes, the two blips moved unerringly toward each other—then merged.

Pieces of the United DC-8 were strewn over the entire width of a street in Brooklyn. The DC-8 crashed there after colliding with a TWA Constellation.

The time was 10:33 A.M.

The Connie crashed onto a field

One of the DC-8's outboard engines sliced into the Constellation's fuselage like a giant meat cleaver. The stricken smaller plane fell onto a field only 100 yards from a public school filled with children.

DC-8 tore up a Brooklyn street

The crippled jet wobbled toward the earth over the borough of Brooklyn, where it yielded to the inevitable. One wing tilted and hit the roof of a four-story apartment building. The big jetliner bounced onto a street, cartwheeled, and smashed into the Methodist Pillar of Fire Church. Aircraft and house of worship exploded into flames. Burning debris spewed in every direction and touched off ten more building fires. The DC-8 jet finally crashed on Sterling Place, a street only one block away from two parochial schools occupied by 1,700 students and teachers. The only part of the plane left intact was the tail section with the United logo plainly visible on the fin. Pieces were strewn over the entire width of the street in Brooklyn.

The initial death toll was 133, including 6 bystanders caught in the open when the jet came down. Of the 128 passengers and crew members aboard the two planes, only 1 survived. The 134th fatality was an 11-year-old boy who had been sitting in the rear of the DC-8. He was pulled out of the flaming wreckage, miraculously alive. But he suffered burns over 80 percent of his body and died the next day.

FAA spends $500 million to improve air control

People all across the nation demanded answers to some tough questions. Great amounts of money had just been spent to improve air safety. The interest in safety came in the wake of a collision in 1956 between TWA and United aircraft over the Grand Canyon—in which 128 people died. At that time the Federal Aviation Administration (FAA, formerly the Civil Aeronautics Administration) poured nearly $500 million into various air traffic control projects. As recently as a month before this latest collision occurred, the FAA administrator, Elwood "Pete" Quesada, proudly told a National Press Club audience in Washington that jets were being monitored by radar "from takeoff to touchdown."

If that were true, then what had gone wrong at 5,000 feet over Staten Island, New York?

Details of the Crash

FAA administrator Quesada's boast of "takeoff to touchdown" radar monitoring turned out to be high-flown exaggeration. Testimony and tape recordings of air traffic control communications revealed the shocking truth.

Only 88 seconds before the collision, Idlewild traffic control announced to United, "Radar service terminated." But at that moment the DC-8 was already far off course. Air control should have noticed the glaring discrepancy and immediately warned United that it was not "approaching Preston" but already had passed it.

Two minutes before the two craft collided, the air control center at Idlewild (later renamed Kennedy International) monitored the United jet on radar. It advised United that if it had to hold at Preston, New Jersey, a radio checkpoint, it was to execute the usual "racetrack" pattern for circling at an assigned altitude until cleared for final approach. United confirmed this instruction.

One minute later, the pilot told Idlewild traffic control that the jet was descending from 6,000 feet to 5,000 feet. Traffic control acknowledged and added these key words:

"*Radar service is terminated* [italics added]. Contact Idlewild approach control. Good day."

United immediately switched radio frequencies and informed Idlewild approach control, "Approaching Preston at five thousand."

Approach control advised United to maintain 5,000 feet, reported that no traffic delays were expected, and provided the flight with the current airport weather conditions along with landing instructions. That communication took only 16 seconds to complete, but the United flight crew could not acknowledge it—it had already collided with the TWA Connie.

United DC-8 was off course

In order for United to fly in only 16 seconds from the Preston checkpoint, where it said (or thought) it was, to Staten Island, where it collided, the jet would have had to be flying at speeds of more than 1,000 miles per hour. And this impossible timetable established two facts:

- The DC-8 jet was not where its instruments apparently indicated it was. At the moment that United radioed "approaching Preston," the jet actually was *11 miles beyond Preston,* racing at nearly 500 miles per hour through the buffer zone that separated Idlewild's traffic from LaGuardia's. Both figuratively and literally, it had run an aerial red light.

- Radar could not have been monitoring the United flight very closely. Only 88 seconds before the collision, Idlewild traffic control announced to United, "Radar service terminated." But at that moment the DC-8 was already far off course. Air control should have noticed the glaring discrepancy and immediately warned United that it was not "approaching Preston" but had already passed it.

A key navigation control malfunctions

The Civil Aeronautics Board (CAB) investigated the incident. They revealed that one of the DC-8's two key navigation instruments was not operating, and United's crew failed to report this condition to Idlewild traffic control. In the board's final report, the CAB put the blame entirely on United and listed the faulty equipment as the accident's probable cause. CAB further claimed that if Idlewild traffic control had known of the malfunction, they would have handled the DC-8 flight differently.

The CAB's announcement was one of the most controversial and angrily challenged accident reports ever handed down, as subsequent events demonstrated.

Impact

The FAA, headed by Pete Quesada, accepted the board's report blaming United and its two dead pilots. Now that blame was established, people could seek legal restitution. Wrongful death claims brought by 115 individual lawsuits sought $77 million in damages. Codefendants named in the suits included not only United and TWA, but Quesada's own agency.

Two government agencies bicker

Quesada protested that neither the FAA's air traffic control system nor any individual controller was in the least way responsible, and he maintained that the indictment of United was eminently fair. Yet his eager acceptance of the aeronautic board's finding was ironic, for the feisty FAA chief had been feuding bitterly with the board. Quesada believed the FAA, not the CAB's Bureau of Safety, should have sole authority to investigate air crashes. But if the FAA were to manage accident investigations, the FAA, as a regulatory agency, might one day end up investigating itself.

The bureau, in turn, resented Quesada's practice of showing up occasionally at the scene of major accidents (as he did after the Staten Island collision, which resulted in a shouting match with a CAB official). They also disliked the way he speculated to the media about the causes of accidents not under his jurisdiction.

FAA orders reduced speeds near airports

Though Quesada loudly denied that his controllers had any blame in the collision, he implemented changes in the air traffic control system. Acting on his orders within days after the collision, the FAA drastically reduced speed limits for aircraft entering a terminal area. It began assigning extra controllers at high-density traffic centers, their specific duty being to watch radar scopes for any planes straying from terminal clearances and holding patterns. And the agency also put into effect improved radar hand-off protection, the process under which one air traffic control center hands off traffic to another. These actions seemed to be an indirect admission of involvement, a point not lost on the public—and the lawsuits came pouring in.

By the time the legal wars ended, Quesada no longer headed the FAA. After five months of hearings before a federal judge, with the testimony filling some 14,000 transcribed pages, the two airlines and the Justice Department—representing the FAA—agreed to share responsibility to varying extent: United Airlines accepted 60 percent of the liability, TWA 15 percent, and the federal government 25 percent.

New FAA chief Halaby vigorously defends air controllers

Quesada's successor, Najeeb Halaby, fought the compromise settlement—in vain—although he managed to take his case all the way up to the attorney general, Robert F. Kennedy. Halaby argued that it was a slap in the face to controllers—"I didn't want our men to carry the stigma of implied guilt when they were so sure in their own minds that they weren't guilty," he explained later.

"I thought the compromise ignored the true role of the controller," Halaby wrote in his autobiography *Crosswinds*. "The term 'controller' implies authority to dictate to flights. But a controller doesn't really control; he *co-operates*. A pilot . . . can ignore advisories and warnings, just as the TWA crew did when Idlewild traffic control warned them that radar had spotted approaching traffic."

Halaby's defense of his men was in character; he enjoyed greater popularity among controllers than any FAA administrator before or since his tenure. But long after he left government service, he admitted to a journalist friend that the warnings issued to TWA were futile anyway—in an overcast sky, the prop-driven Connie was helpless to take evasive action against a fast-approaching jet the crew couldn't even see.

United-TWA collision pointed out need for further safety measures

Yet it also seems fair to blame the Staten Island crash on the system, if not individual controllers. The 1956 Grand Canyon collision—ironically involving the same two airlines—spurred the FAA to spend millions on what amounted to multilane aerial turnpikes. While this was in itself a massive and worthwhile achievement, it was not enough to prevent the 1960 collision from happening. Despite improved routing between the nation's airports, not nearly enough had been done about the one-lane traffic at the end of these airways.

Where to Learn More

"Civil Defense." *Nation* (December 31, 1960): 513–514.

"Crash Stirs New York Air Traffic Probe." *Aviation Weekly* (December 26, 1960): 27–29.

"Death in the Air: Collision in New York." *Time* (December 26, 1960): 14–15.

Hoffman, D. H. "CAB Probe Centers on DC-8 Recorder." *Aviation Weekly* (January 16, 1961): 38–39.

"New York Collision Hearings Point to New CAB, FAA Friction." *Aviation Weekly* (January 9, 1961): 31.

Ruppenthal, K. M. "Crash of a System: New York's Air Tragedy." *Nation* (December 31, 1960): 516–519.

American Airlines DC-10 crashes

Chicago, Illinois
May 25, 1979

When the left wing's engine rips away from the DC-10 during takeoff, the ensuing crash kills 273 people and prompts the FAA to cancel the airworthiness certification of the DC-10.

Background

On May 25, 1979, as American Airlines Flight 191 took off from Chicago's O'Hare International Airport, engine Number 1 on the DC-10 broke away from the left wing. Though heavily loaded for its trip to Los Angeles, the aircraft managed to climb to 325 feet, then rolled to the left into a fatal dive.

Fiery crash claims 273 lives

Flight 191 cut a swath in the ground with its left wing and narrowly missed three large fuel storage tanks before it crashed, less than a mile from the end the runway. Then it exploded. Eyewitnesses compared the fireball to a nuclear "mushroom cloud." All 271 persons onboard the DC-10 and 2 persons on the ground died in the crash. It was American Airlines' worst aircraft accident.

However, this was not the DC-10's first major accident, nor would it be the last. Other DC-10s built by McDonnell-Douglas had succumbed to major disasters. In 1974 a Turkish Airlines DC-10 crashed during takeoff from Paris-Orly Field after a cargo door blew out. The resulting gap in the side of the fuselage depressurized the area underneath the cabin floor, which caused the floor to collapse and sever the aircraft's control cables. The DC-10 crashed after a free-fall dive from 14,000 feet, and 346 people perished. Investigators blamed the crash on a poorly designed cargo-door latch.

McDonnell-Douglas casually "recommends" a major repair

However, most of the DC-10's problems concerned pylons. Pylons

Firemen examine the smoldering crash site of Flight 191. The DC-10 plunged to the ground after its left engine fell off the wing; the engine's rear attachment bulkhead and flanges had been damaged during maintenance procedures.

attach the underwing engines to the wing itself. (The DC-10 has three engines, one on the tail and one under each wing.) McDonnell-Douglas called for the replacement of some of the upper and lower bearings on the pylons and issued service bulletins regarding the pylons in May 1975 and again in February 1978. Service bulletins are recommendations from the manufacturer about how a specific maintenance procedure should be handled.

A pylon weighs over 1,800 pounds, and combined engine and pylon weight is 13,477 pounds (more than 6 metric tons). Replacing the bearings would thus be a major repair procedure that could only be attempted at well-equipped maintenance facilities. McDonnell-Douglas recommended that aircraft operators change these bearings when an engine needed to be removed or replaced, since this larger procedure would expose and allow easy access to the pylons attached to the wing. Accordingly all American commercial airlines operating the DC-10 at the time—American, United, Continental, and National—scheduled pylon bearings replacement to be implemented at the same time as routine engine dismounting occurred.

American uses forklift during repairs—over manufacturer reservations

American Airlines engineering and maintenance personnel had to develop instructions for their maintenance technicians, called an internal Engineering Change Order (ECO). Logically they sought the most efficient approach to implement McDonnell-Douglas' recommendations and assessed the feasibility of using a large forklift. Using a forklift to remove the entire engine/pylon package would reduce the man-hours and costs involved. American Airlines believed it could save over 200 man-hours per engine and reduce the number of hydraulic and electric line connections and disconnections necessary.

American Airlines knew that its competitor, United Airlines, was using an overhead crane to remove the engine/pylon package from its DC-10s, so American contacted McDonnell-Douglas to discuss American's approach to the procedure. The engineering staff at McDonnell-Douglas expressed reservations about the proposed method and responded that "Douglas would not encourage this procedure due to the element of risk involved in remating the combined engine and pylon assembly to the wing." However, since American itself owned the aircraft, McDonnell-Douglas did supply recommendations.

The manufacturer advised that during the lowering and raising of the engine/pylon package, the forklift should lift the engine assembly precisely at its center of gravity (CG) to reduce the risk of damage while the engine was being supported by the forklift. According to the manufacturer's recommendation, American Airlines engineering personnel calculated the CG of the engine/pylon package, but maintenance technicians did not abide by the mathematical CG. Instead they positioned the forklift under the engine assembly on the basis of a visual estimate.

Supporting 6 metric tons, the forklift strains and its load shifts

While the engine was being lowered and raised, the assembly frequently shifted slightly. The American Airlines maintenance record indicated that the forklift lowered slightly under load (when holding up the engine assembly), which would subject any portion of the engine still attached to the aircraft to great stress. Moreover, because of the close fit of the pylon to the wing and the minimal clearance between structural ele-

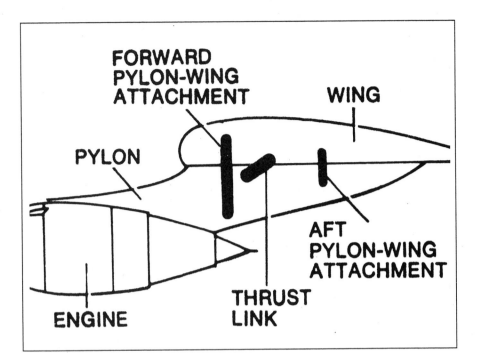

This diagram shows the initial takeoff position of the left-wing pylon-engine assembly.

ments, American's maintenance personnel had to be extremely careful when performing the dismounting procedure. A minor mistake or movement by the forklift operator could easily damage the rear attachment bulkhead and flanges. In the case of the DC-10 aircraft used for Flight 191, this is precisely what happened.

Details of the Crash

At Chicago's O'Hare International Airport on May 25, 1979, just after 3:00 P.M., Flight 191 began its takeoff roll down Runway 32 Right. The pilots were unaware of any problem with the left (Number 1) engine, because the engines on the DC-10 are located on the underside of the wings and are not visible from the cockpit. Then, just before liftoff from the runway, engine Number 1 broke away from the aircraft, trailing behind the pylon designed to connect it to the wing. As the left engine ripped off the wing, it tore away three feet of the wing's leading edge.

With stall warning disabled, pilots choose wrong corrective action

When the pilots realized they had a problem, there was not enough runway left to brake the aircraft. Emergency procedures under such circumstances call for the pilots to complete the takeoff. Loss of one engine meant that the aircraft's performance would be reduced, but the pilots expected to be able to fly on the two remaining engines, even considering that the plane was heavily loaded. However, the procedures the pilots relied on were based on power failure in an engine, not on the engine physically breaking away from the aircraft.

When engine Number 1 separated from the aircraft, it tore loose all associated hydraulic and electrical lines. With hydraulic pressure gone, control of the slats—devices to increase lift—on the wing's leading edge was lost. Normally hydraulic pressure lowers the slats for takeoff, but instead the slats retracted back into the wing. The backup system for holding these slats down also failed, because it too relied on the hydraulic lines. With the slats retracted into the left wing during takeoff, the wing would stall, or stop "flying," at a much higher airspeed than normal. With engine Number 1 missing, the aircraft could not maintain speed during the takeoff climb. The aircraft began to decelerate and approach the stall speed of the left wing.

Simulations prove the accident could have been averted

If aircraft instrumentation signals an approaching stall, the crew can make appropriate maneuvers to avert the stall, but with the electrical system damaged by the fallen engine, the stall warning system was also inoperative. Electrical power was dependent on the generator in engine Number 1. There was no backup system on the DC-10 to override this situation. The pilots were never aware that the wing was about to stall. If they knew, they might have averted the crash, according to post-crash tests conducted in a DC-10 simulator. Recreating the last moments of Flight 191, cockpit crews who knew that their stall warning device was inoperative were able to lower the nose of the DC-10, keep the aircraft flying, and prepare for a landing.

Unaware of their own situation, the crew of Flight 191 tried to keep the DC-10 climbing, but its airspeed decreased and reached the 159-knot stalling speed of the wing. Then the aircraft began a slow roll to the left that the pilots were unable to control. When the aircraft rotated past the point where the wings were perpendicular to the ground, the nose of the DC-10 dropped, and Flight 191 began its final descent.

Impact

In the accident report regarding the crash of Flight 191, the National Transportation Safety Board (NTSB) exonerated the pilots of any responsibility for the accident due to the various system failures that had occurred. The NTSB report blamed the crash on the maintenance procedures devised by American Airlines and on the design of the DC-10 engine pylons.

Repair procedure is indicted for causing accident

The NTSB determined that overstress cracks in the flanges in the rear of the engine pylons precipitated the disaster. The overstress cracks resulted from American Airlines' method of engine/pylon removal and the stresses exerted by movements of the forklift. Subsequent tests performed by both American Airlines and McDonnell-Douglas confirmed that a movement of the forklift as slight as six-tenths of an inch could damage the engine attachment flanges and bulkheads. Continental Airlines DC-10s undergoing maintenance on engine pylons on December 19, 1978, and on February 22, 1979, had been damaged in exactly the same way. Continental Airlines did not report this damage to the Federal Aviation Administration (FAA), however, nor were they required to do so.

The NTSB was very pointed in its comments on the design of the DC-10 engine pylons. The engine pylon design relied on attachment devices that were susceptible to damage during maintenance procedures which in any event were inherently difficult to perform. The NTSB maintained that "McDonnell-Douglas should have foreseen that pylons would be removed, and therefore the parts of the rear bulkhead [the flanges] should have been designed to eliminate, or at least minimize, vulnerability to damage during maintenance."

FAA grounds fleet from entire U.S. airspace until DC-10s can be repaired

The design of the hydraulic and electrical flight control systems also concerned the NTSB. The loss of these systems kept the cockpit crew on Flight 191 from determining the nature or extent of their problem. Although the original certification of the DC-10 required the aircraft to be operable in cases of failure of certain lift devices, the regulation applied only to the wings' trailing edge flaps, not to the leading edge slats, which in the case of Flight 191 retracted and were completely inoperable. NTSB investigators confirmed that engine and slat failure jeopardizes the ability of the DC-10 to climb safely. The NTSB acknowledged finally that although DC-10 certification complied with all current regulations, the regulations themselves may have been inadequate and may have contributed to the Flight 191 crash.

The NTSB advised McDonnell-Douglas to provide to all airlines flying DC-10s some emergency instructions about the improper maintenance procedures that caused the American Airlines accident. The FAA took the NTSB's recommendations one step further and on June 6, 1979, revoked the airworthiness certificate of all DC-10 models. The FAA pronouncement for all practical purposes grounded the American commercial aircraft fleet until repairs of the pylons could be completed and maintenance procedures corrected. On June 26, 1979, the FAA broadened their restrictions on the DC-10 through a Special Federal Aviation Regulation that "prohibited the operation of any Model DC-10 aircraft within the airspace of the United States." Foreign airlines, therefore, were also prohibited from using the aircraft to land in or depart from any U.S. airfield. After the FAA confirmed that the airlines had performed the necessary repairs and updated the maintenance procedures, the agency on July 13, 1979, recertified the McDonnell-Douglas DC-10 as airworthy.

Where to Learn More

"Disaster on a Jumbo Jet: Crash of an American Airlines DC-10 at Chicago's O'Hare International Airport." *U.S. News and World Report* (June 4, 1979): 14.

North, David M. "Crash to Boost FAA Scrutiny." *Aviation Week and Space Technology* (June 4, 1979): 12–15.

Williams, Dennis A., and others. "It Just Disintegrated." *Newsweek* (June 4, 1979): 24ff.

Witkin, Richard. "U.S. Halts DC-10's Indefinitely, Demanding New Safety Tests: Travel Is Disrupted for 60,000." *New York Times* (June 7, 1979): 1.

"Worst U.S. Air Crash." *Time* (June 4, 1979): 12–13.

Japan Airlines Boeing 747 crashes

Remote mountain forest in Gumma, Japan
August 12, 1985

Background

On August 12, 1985, the eve of the three-day celebration of Bon (Feast of Lanterns), Japan Airlines' (JAL) Flight 123 crashed into the side of Mount Osutaka in remote Gumma Prefecture, killing 520 of 524 people aboard. The fully booked aircraft was on its way to Osaka. It had left Tokyo's Haneda Airport just 55 minutes earlier. The disaster led to improved worldwide inspection schedules. The investigation also reaffirmed the basic integrity of Boeing 747 aircraft design.

The aircraft registered as JA8119 was a Boeing 747SR-100. (The SR designation means "short-range.") Boeing manufactured and delivered it to JAL in 1974. The short-range version of Boeing's wide-body transport was developed specifically for Japan's domestic air route system, which is characterized by high-density, short-distance routes. Transport serving this kind of market must withstand the rigors of frequent takeoffs and landings for relatively brief flights with large passenger loads.

Crash of a Boeing 747 aircraft, which kills 520 people, is blamed on metal fatigue caused by faulty repairs.

JA8119 was damaged in a 1978 rough landing

Short-range operations mean that a greater number of takeoff and landing cycles occur (as well as their accompanying pressurizations and depressurizations). These cycles age an aircraft and cause certain parts to fatigue more rapidly. The dimensions and external appearance of the JA8119 were similar to the basic 747 version, but Boeing added over 3,000 pounds of structural material. The aim was to strengthen the 747SR to withstand the additional stresses imposed by flying domestically and to assure a "fatigue lifetime" equivalent to other 747 models. In a 20-year

Flight 123 crashed in a remote mountain forest; it took firemen over 14 hours to reach the area. The accident was blamed on metal fatigue resulting from faulty repairs.

lifetime of short-range service, a 747SR is expected to experience 52,000 landings. At the time that the 11-year-old JA8119 departed on its ill-fated takeoff, the aircraft had flown over 25,000 hours and logged nearly 19,000 cycles since delivery.

Yet there was a repair history on this jet that would later prove critical to the crash of Flight 123. Seven years earlier JA8119 suffered damage during a hard landing at Osaka International Airport. On June 2, 1978, the aircraft sent up a shower of sparks as it dragged its tail on the runway. The airframe was substantially damaged and 30 people on board were injured. After provisional repairs were made by JAL at Osaka, the aircraft was ferried to Tokyo for complete repairs to be carried out by a Boeing repair team.

Extensive repairs grounded JA8119 for two weeks

It was normal procedure for repairs to be contracted out to Boeing. The jet's manufacturer had a satisfactory repair record. Boeing reportedly replaced 54 feet of skin under the lower fuselage. Then it riveted a new lower half of the rear pressure bulkhead to the original upper half. The rear pressure bulkhead, located near the lavatories at the back of the passenger cabin, is made of aluminum alloy. Shaped like an umbrella canopy, the bulkhead is a thin partition that separates the passenger cabin, which is highly pressurized in flight, from the unpressurized tail cone in the rear (the stern).

Repairs grounded JA8119 for two weeks. When the aircraft was back in operation, a JAL spokesperson declared, "As far as we are concerned it [is] as good as new." But flight attendants reported hearing unusual noises in the stern, and over time they worsened. Still, the aircraft accumulated another 16,000 flying hours and more than 12,000 landings since the Osaka incident, and it managed to pass two prescribed maintenance inspections.

Repaired area is susceptible to fatigue

It came to light later that an error occurred during the repair work. A small part of the bulkhead splice—where the new and old components of the pressure bulkhead overlapped—should have been joined using two rows of rivets. Instead, the splice received only a single row, decreasing the strength of the splice to about 70 percent of its original strength. This left the "repaired" area susceptible to fatigue.

Details of the Crash

The aircraft landed at Tokyo International Airport from Fukuoka. There was less than an hour before it was due to make the Flight 123 run at 6:00 P.M. to Osaka International Airport. JA8119 was inspected in preparation for the 250-mile, one-hour flight. During this run the copilot, who was being trained for the position of captain, switched to the left seat. The captain took the right chair.

Loud boom heard 15 minutes into flight

The takeoff was normal. Some fifteen minutes later the cockpit made a routing request while the aircraft climbed to its cruising altitude of

The aircraft continued a rapid descent eastward. At about 6:45 P.M. the crew repeated that the plane was "uncontrollable." Indeed it was flying wildly between altitudes of 7,000 feet and 13,000 feet. At 6:55 P.M. the cockpit acknowledged a transmission from Haneda and Yokota U.S. Air Force air base approach controls, informing the jet that it could make an emergency landing at either place. But the crew made no further communications.

24,000 feet. At that altitude, an unusual vibration raised the nose of the aircraft and caused a loss of rudder control. A loud boom followed, and the cockpit immediately signaled an emergency to the tower at Tokyo.

The captain requested clearance to descend to and maintain 22,000 feet of altitude and to return to Haneda. The ground crew responded with route instructions. Suddenly the plane began to "dutch roll," its nose swaying uncontrollably left and right. It also began to suffer altitude and speed changes known as phugoid oscillations, which would continue until just before the crash.

JA8119 "now uncontrollable"

When ground asked the nature of the emergency, the aircraft did not respond. Tokyo repeated its route instructions but at about 6:30 P.M. received the response "now uncontrollable." In the next two minutes the aircraft changed course northbound, responded "now descending," and reported an altitude of 24,000 feet.

Although Flight 123 was only 72 nautical miles from Nagoya Airport, the cockpit requested to return to Haneda. It veered right and left near Mount Fuji. Then, around 6:40 P.M. the aircraft descended from 21,000 feet over Otsuki City to about 17,000 feet, flying a two-minute circle in three minutes.

Aircraft totally disintegrated on impact

The aircraft continued a rapid descent eastward. At about 6:45 P.M. the crew repeated that the plane was "uncontrollable." Indeed it was flying wildly between altitudes of 7,000 feet and 13,000 feet. At 6:55 P.M. the cockpit acknowledged a transmission from Haneda and Yokota U.S. Air Force air base approach controls, informing the jet that it could make an emergency landing at either place. But the crew made no further communications.

The ensuing events were pieced together only after the disaster, which cost the lives of 520 of the 524 people aboard Flight 123. Eyewitness statements from the four survivors indicated that the jet made an abrupt right turn before it suddenly plunged into a dive, banking to the left behind Mount Mikuni. The aircraft hit the top of a hill with its right wing. It sheared several trees and lost an engine. The aircraft then crossed a ravine, where it lost another engine. Then it hobbled 656 yards forward and struck the top of a 6,233-foot peak. JA8119 exploded at about 6:56 P.M. and almost totally disintegrated on impact. Sections catapulted into ravines on either side. Because Flight 123 crashed in such a remote mountain forest, even helicopters could not land safely. It took firemen over 14 hours after the crash to reach the area.

Japan's Ministry of Transport informed the Aircraft Accident Investigation Commission to begin the investigation and appoint an investigator-in-charge. Assisting the commission were members of Japan's Ministry of Transport, the U.S. Federal Aviation Administration (FAA), and teams from the U.S. National Transportation Safety Board (NTSB) and Boeing.

Japan's traditional reverence for the dead placed a high priority on locating and recovering the bodies of the 520 who died in the crash. This meant cutting and moving pieces of wreckage that could be critical to the investigation, including the rear pressure bulkhead. The first clues to what caused the fatal crash—parts of the tail fin and lower rudder—were found in Sagami Bay. Almost half the tail fin detached in flight—investigators were amazed the aircraft stayed in the air as long as it had.

Investigation focuses on the single line of rivets

Examination of the bulkhead revealed a troubling fact. After sustaining extensive damage during the rough 1978 landing at Osaka, a single line of rivets fastened part of the repair when the repair manual specified a double line. The faulty repairs allowed abnormal weakening of the bulkhead, which then could not contain the air pressure in the cabin. As a result the bulkhead blew out in flight. This caused a series of succeeding ruptures, like a chain reaction. The rapid ventilation of pressurized air into the tail section destroyed the vertical tail fin. It also destroyed all four hydraulic systems, which are vital to aircraft control. Loss of JA8119's hydraulics explains why the aircraft careened wildly out of control.

Impact

Profound shock affected the Japanese nation. There were 509 passengers (including 12 infants) and 15 crew members on board. Of these, only 4 passengers survived, and they sustained severe injuries. All 4 survivors were seated together in the center of row 56. One of them was an off-duty JAL stewardess. In a gesture of corporate "guilt," Japanese Airlines president Yasumoto Takagi resigned, shouldering full responsibility. Shamed by the high death toll, a JAL maintenance manager at Haneda took his own life in apology for the disaster.

Was the 747 unsafe?

Summer 1985 was a dark time in aviation history. Another 747, Air India Flight 182, crashed, just two months apart from Flight 123. Due apparently to some kind of structural damage, the crash killed 329 people. The aviation world was deeply alarmed by the events. Suddenly it seemed that the 747 might be structurally unsafe. Everyone wanted answers.

Boeing and JAL set up a joint fund on a 50/50 basis to compensate victims of the disaster—although neither company claimed to be liable. When the investigation on the Air India disaster concluded (the tail section was found intact), there was no link to JAL's JA8119. Flight 123 succumbed, instead, to improper repair. Furthermore, regular inspections failed to detect the cracks in the bulkhead. Boeing officials and aviation authorities were relieved, because there were more than 600 747s in service. They had been fearing that some generic defect in the 747's design might repeat the Japanese disaster. Establishing the cause as a faulty repair dispelled fears. And to find any similar repair defects before accidents occurred, inspection programs were modified.

Faulty repairs on JA8119 were difficult to detect

As a result of this accident the design of the 747's empennage (tail assembly) was modified. The goal is to protect control surfaces from failing catastrophically if sudden depressurization occurs. The vent door in the 747's original design did not protect adequately if a large area of the bulkhead opened suddenly. Installing a structural cover for the opening within the empennage produced the desired protective effect and still allowed for access to the vertical fin. The design of the hydraulic systems was also modified to prevent the simultaneous loss of all four systems in case of pressure buildup.

Repair procedures of rear pressure bulkheads were also reevaluated to ensure that they would reinforce Boeing's fail-safe concept. The idea behind the initial design of the rear pressure bulkhead was the so-called one-bay fail-safe. Engineers assumed that a crack in the bay (or compartment) could be detected and repaired while remaining within that bay—a single area surrounded by stiffeners and tear straps. They never considered the situation of cracks being present simultaneously in several bays, which proved to be the case with JA8119. Nevertheless, it was difficult to detect the cracks visually and even by air leakage. Repairs were covered with thick sealants, and the cracks were small and spread along rivet holes at web overlaps.

In cases of doubt about the validity of repair records, government inspectors were urged to check rear bulkheads. They set up a long-range program monitoring damaged airframe structures. They revised the 747's rear pressure bulkhead inspection program. This change provides inspections beyond routine visual inspection to detect the extent of possible multiple-site fatigue cracks. Boeing also called for a switch to more corrosion-resistant bolts in the empennage.

Cracks discovered in forward sections of 747s; worldwide inspection ordered

In January 1986 JAL engineers discovered cracked and broken ribs in the frames of the forward sections of some 747 fuselages. Other airlines reported the same cracks. The FAA issued an emergency directive requiring worldwide inspection of the fleet, and Boeing recommended internal visual inspection. Detailed aircraft inspection and prompt repair work overcame the frame-cracking problem—it also outlined the success of structural monitoring.

The JAL crash was blamed on maintenance error. No other incriminating evidence turned up regarding the 747's basic design, manufacture, or structural integrity, which confirmed the airworthiness of the aircraft.

Thirty-three damage suits (Japan and the United States combined) were filed against JAL as a result of the 1985 disaster. All cases were settled out of court, with compensation averaging approximately $850,000 per person. In 1993 the Japanese Ministry of Transport removed the 1929 ceiling on compensation for victims of international flights. (The ceiling for victims of domestic flights had been unlimited since 1982.) This helps victims of air disasters obtain compensation more swiftly by eliminating the need for litigation.

Where to Learn More

Aircraft Accident Investigation Commission, Ministry of Transport (Japan). "Aircraft Accident Investigation Report." Trans. L. A. Turner. *JAL JA8ll9: August 12, 1985.* Aircraft Accident Investigation Commission, Ministry of Transport (Japan), June 19, 1987.

Allen, Glen. "The Fatal End of JAL Flight 123." *Maclean's* (August 26, 1985): 42.

Anderson, Harry. "What Went Wrong?" *Newsweek* (August 26, 1985): 14–17.

ICAO. "Japanese Civil Aircraft Accident Interim Report." *Report No. 1985-6 (JAL).* August 1985.

"Japan Orders Checks of 747 Tail Sections after JAL Crash." *Aviation Week and Space Technology* (August 19, 1985): 30.

"Japanese Cite Faulty Bulkhead Repair as Cause of JAL 747 Crash in 1985." *Aviation Week and Space Technology* (June 29, 1987): 34.

Haberman, Clyde. "Part of Jet's Tail Is Found 80 Miles from Crash Scene." *New York Times* (August 14, 1985): 1.

O'Lone, Richard G. "747/767 Pressure Bulkhead Tests Precede Suggested Safety Changes." *Aviation Week and Space Technology* (December 16, 1985): 29.

Witkin, Richard. "Clues Are Found in Japan Air Crash." *New York Times* (September 6, 1985): 7.

———. "Crash Mystery: What Damaged Tail?" *New York Times* (August 14, 1985): 8.

United Airlines Boeing 747 explodes

Over the Pacific Ocean 100 miles south of Honolulu, Hawaii
February 24, 1989

Background

On February 24, 1989, about 18 minutes after takeoff from Honolulu, Hawaii, a United Airlines 747-100 passenger jet bound for Auckland, New Zealand, experienced explosive decompression. The forward cargo door opened, which disrupted the even distribution of pressure in the airplane's cabin and cargo hold. The leak precipitated a massive rush of air that punched a gaping hole in the fuselage above the door and vanished nine business-class passengers out the cavity.

Stray electrical signals defeat lock and latch system

The subsequent accident investigation dragged on for three years. It ultimately concluded that stray electrical signals could defeat an elaborate system of latches and locks. Worse, the 747's cargo door could spontaneously open itself up and precipitate cabin decompression. Extensive improvements in the 747's cargo door system were developed in the aftermath of the disaster.

The Boeing 747 jumbo jet was designed in the late 1980s. Recognized by its distinctive humped fuselage, the 747 is one of the most successful jetliners ever built. More than 900 747s are in commercial service, and production is scheduled to continue into the foreseeable future. Each 747 has two cargo doors, measuring nine square feet in size, located on the starboard (right) side of the airplane's belly, one forward and one aft. These cargo doors differ markedly from the customary inward-opening "plug doors," which wedge tightly into the passageway as the aircraft pressurizes. The 747 cargo doors swing out and up, gull-wing fashion.

Explosive decompression blows a giant hole in the 747's side and vanished nine passengers. The cargo doors' electromechanical latches and locks succumbed to stray electrical signals.

This photograph shows the damage on the starboard side of the fuselage. At the moment of explosive decompression, passengers and cabin attendants reported a hurricane-force blast of cold air.

Successful cargo doors develop troubling cluster of problems

Plug doors were used almost exclusively to admit passengers and cargo on commercial jetliners—until the 747 came along. Plug doors were considered a necessary safety measure while the airplane is pressurized, should the integrity of the fuselage ever be breached. Designers of the 747 took a different approach, because they wanted to avoid the heavy tracks and wide inside clearances—wasted space—required to accommodate the giant plug doors.

So they designed a relatively light door (about 800 pounds) that swung out and up and did not obstruct any cargo space. The new design would still be subject to the same enormous pressurization forces. To assure that the door would always hold fast against them, they devised an intricate system of electromechanical latches and locks. The system permits a ground worker to lower and shut a door in about 15 seconds by depressing a toggle switch that engages a series of electrical motors. As a final step the worker locks the door manually by depressing a handle in the middle of the door. The 747's outward-opening cargo door proved so successful that it became the standard for the next generation of jetliners.

Yet soon after the 747 entered commercial service in the early 1970s, a vexing trend emerged. One by one the door's complicated system of sequential electronic actuators began to fail:

- Locks became battered and bent
- Clutches and motors were stripped
- Sensors and switches burned out
- Latches and locks jammed.

Forward cargo door opens 1.5 inches during flight

Boeing recommended to the airline companies ways to correct each problem. Then on March 10, 1987, a Pan Am 747-100 was about to depart London for San Francisco. Its forward cargo door was lowered and shut—manually—by a ramp worker using a speed wrench. This was considered a routine procedure to be used when the door's electrical system malfunctioned. Shortly after takeoff the pilot found he was unable to pressurize the cabin, so he returned to London. There the worker discovered that the forward cargo door was open 1.5 inches. The door was closed and locked and the flight was resumed.

A closer inspection was conducted when the 747 reached San Francisco. Eight boomerang-shaped door locks, which are designed to secure eight door latches in a closed position during flight, were severely damaged. When the latches work properly, they engage and then rotate to a closed position around latch pins, which are fastened to the door sill. The locks then swing into place over the open end of the latches, blocking them from reverse-rotating to the open position.

Ground crew worker blamed, yet latch problems are seen fleet-wide

Boeing conducted lab tests afterwards. The switch is designed to cut all electrical power from the door once the locks are set, but engineers discovered that the switch could jam. This left the door's mechanisms vulnerable to stray electrical signals that could actuate the latch motor. The motor could then "backdrive" the latches to an open position, bending the locks out of the way in the process. Tests showed that if the latch motor were somehow activated, the locks—made of aluminum—were too weak to prevent the latches from rotating open.

Testing also showed that a stray electrical signal could almost instantly backwind the latches past weak locks to a fully open position. An inspection of Pan Am's 747 fleet revealed numerous damaged locks, suggesting a chronic pattern of weak locks being frequently subjected to latch backwinding.

Meanwhile, officials blamed the Pan Am incident on the ground worker who manually operated the door. Though the worker denied doing this, he was said to have used his speed wrench to backwind the latches after setting the locks in place. Officials theorized that this was a shortcut procedure devised by ground workers to confirm that the latches were snugly secured against the locks. The mechanic would have had to make 95 full turns of his speed wrench to fully open the latches. The latch motor accomplishes the same task in 15 seconds.

Boeing warns airlines of rapid decompression

About one month after the Pan Am incident, Boeing alerted airlines worldwide that opening of a cargo door in flight "could result in rapid decompression . . . resulting in collapse of the passenger cabin floor and the possible damage to airplane electrical and hydraulic systems." The manufacturer called for airlines to reinforce the weak aluminum locks with special steel braces. Pan Am immediately fabricated the braces and installed them on its 747 fleet. Meanwhile, the Federal Aviation Administration (FAA) considered it a low-priority matter and gave airlines up to two years to make the upgrade.

Details of the Explosion

In December 1988, the maintenance logs of the United Airlines jumbo jet that flew the Honolulu-Auckland route documented the chronic problem: the electrical system of its forward cargo door had malfunctioned repeatedly. It had to be operated manually on 14 separate occasions, but there had been no "write-up" on that cargo door after December 1988.

In the early hours of February 24, 1989, routine preparations readied the 747 for Flight 811. The ground crew used the toggle switch to routinely lower and latch the door, and a ground worker noticed nothing unusual when he depressed the lock handle to swing the locks into place over the closed latches. The locks were the weak aluminum type, because United had not yet installed the steel braces.

That day Flight 811 carried 337 passengers, 15 cabin attendants, a pilot, a copilot, and a flight engineer. They were cleared for takeoff from Honolulu, Hawaii, at 1:52 A.M. The jumbo jet was climbing past 22,000 feet at more than 500 miles per hour, and pressurization—pumping compressed air into the cabin and cargo hold—proceeded as usual. While the aircraft was in actuality flying higher than the summit of Mount McKinley, passengers didn't have to rely on breathing thin air; the air in the cabin was pressurized to a comfortable 4,000 feet.

Explosive decompression blows out forward cargo door and nine passengers disappear

At 2:09 A.M. everyone was stunned by a loud thump. Captain David Cronin demanded, "What the hell was that?" A tremendous boom followed 1.8 seconds later—11 million pounds of force distributed on every square inch of the inside of the aircraft gushed explosively toward a breach in the fuselage. As the forward cargo door heaved away, it tore a 13 x 15 foot hole into the fuselage alongside rows 9, 10, 11, and 12 on the starboard side.

In the blink of an eye, business class seats G and H in rows 8 through 12 vanished along with a chunk of the floor, the starboard aisle, and the eight passengers who were sitting in those seats. Seat 9F, bordering the left edge of the starboard aisle, remained on its mounts, but its occupant, a 49-year-old man, disappeared.

Pilot returns crippled jet to Honolulu

At the moment of decompression passengers and cabin attendants experienced a hurricane-force blast of cold air. Cabin attendant Curt

A loud thump was followed by a tremendous boom, the sound of 11 million pounds of force gushing toward a breach in the fuselage. As the forward cargo door heaved away, it tore a 13 x15 foot opening on the fuselage alongside rows 9, 10, 11, and 12 on the starboard side.

Christensen reported: "Immediately the air filled with a hazy smoke and flying debris. I felt a cold wind and sucking, if you will, but it wasn't a sucking. It was like being in the middle of a huge cannon blast, a blast of cold air. There was gray, swirling smoke and debris flying everywhere."

For 20 grueling minutes passengers and cabin attendants—some severely injured—feared that the jetliner would crash into the Pacific Ocean. The two starboard engines flamed out. Miraculously Captain Cronin nursed the crippled jumbo jet to a safe landing back in Honolulu.

Impact

This horrifying disaster convinced the FAA to revise its two-year time frame for the 747 lock upgrade. They ordered the steel braces to be installed within 30 days. Over the next several months, Boeing developed a number of improvements to the cargo door's warning system. As for what caused the accident, officials were stumped. The most telling evidence—the door itself—lay at the bottom of the Pacific Ocean, 100 miles south of Honolulu.

Without that crucial piece of evidence, the National Transportation Safety Board (NTSB) nevertheless ruled on April 16, 1990, that the weak locks probably were severely damaged during the 14 occurrences of manual operation in the two months before the accident. The NTSB decided that the ground crew somehow did not fully close the latches prior to takeoff. This conclusion contradicted the workers' reports that they had no problems using the toggle switch to lower and shut the door and noticed nothing unusual about locking the door.

Navy retrieves cargo door from bottom of ocean

On September 27, 1990, the U.S. Navy deep-submergence vehicle *Sea Cliff* plucked the lower two-thirds of Flight 811's cargo door from the ocean floor. The eight locks were found in the locked position, but bent out of the way by the eight latches, which had somehow reverse-rotated to the open position. All the locks were found in relatively good shape, which contradicted the NTSB's official claim of severely battered locks. A few days later *Sea Cliff* recovered the upper half of the door. Both parts were then shipped to a Boeing laboratory in Seattle, where they underwent thorough analysis under safety board supervision.

Cargo door opens on its own on a newer model 747

While the cargo door was being examined, a newer model United Airlines 747-200 experienced the cargo door phenomenon. On June 21, 1991, at Kennedy International Airport, the 747 was being prepared for a flight to Tokyo. The aft cargo door was closed and latched but not yet locked. Then, while no one was touching the toggle switch, the door latches rotated open and the door lifted up. The powerful door-lift motor continued to run, trying to raise the door past its maximum open position, until a mechanic shut it off by popping a circuit breaker. The stray signal that was found to have caused the event was subsequently traced to a cracked wire-bundle conduit. Located near the door hinges, the conduit was exposed to wear and tear each time the door was opened or closed.

From then on United ordered its ground crews to open a pair of circuit breakers just before pushing any 747 back from the gate.

Accident formally blamed on "faulty switch or wiring"

Combining the evidence of stray signals at work in the door from the United 747 at Kennedy with detailed analysis of Flight 811's cargo door, the National Transportation Safety Board took an unusual step and on March 18, 1992, reversed its earlier ruling. The NTSB formally attributed the opening of Flight 811's cargo door to a "faulty switch or wiring in the door control system which permitted electrical actuation of the door latches toward the unlatched position after initial door closure and before takeoff."

The wire bundle, which would conclusively prove its ruling, was never found, but the NTSB weighed other evidence uncovered by United Airlines. After the Kennedy door-opening incident, the airline detected 21 different simple short circuits that could lead to uncommanded actuation of the latch motor. Boeing believes that the steel reinforced locks now used on all 747s will prevent the latches from moving, even if a stray signal does come into play.

NTSB accused Boeing and FAA of not taking corrective action soon enough

Even so, the NTSB called the design of the cargo door locking mechanisms "deficient" and said that a lack of timely corrective action by Boeing and the FAA after the 1987 Pan Am incident contributed to the tragedy of Flight 811.

Where to Learn More

Acohido, Byron. "Flight 811: Terror in the Sky." *Pacific Magazine* (*Seattle Times/Seattle Post Intelligencer* Sunday supplement) (January 5, 1992): 10–19.

"Airlines Inspect Cargo Doors in Wake of United Accident." *Aviation Week and Space Technology* (March 6, 1989): 22.

"Board Suggests Changing Cargo Door Latch Design." *Aviation Week and Space Technology* (August 28, 1989): 29.

"Close Look at Cargo Door Workings Fails to Reveal Cause of Accident." *Aviation Week and Space Technology* (May 22, 1989): 86.

Hackett, George. "Flight 811: A Nightmare in the Sky." *Newsweek* (March 6, 1989): 26.

Henderson, Breck W. "Investigators Believe Cargo Door Failed, Ripping 747 Open." *Aviation Week and Space Technology* (March 6, 1989): 18.

National Transportation Safety Board. "Explosive Decompression: Loss of Cargo Door in Flight." *Aircraft Accident Report No. NTSB/AAR-92/02*. Washington, DC: Government Printing Office, March 18, 1992, 1–113.

El Al Boeing 747-200 crashes

Amsterdam, The Netherlands
October 4, 1992

Background

On October 4, 1992, after a year-old Boeing 747-200F (freighter) operated by El Al Israel Airlines took off from Amsterdam's Schipol Airport at 6:22 P.M., the inboard engine came loose from the right wing. The engine then veered sideways and knocked off the right outboard engine. The jetliner lurched to the right and, constantly losing altitude, made two wide clockwise circles over the city. The airplane finally smashed into a crowded, mostly low-income ten-story apartment complex, where the residents were just sitting down to dinner. The 6 El Al crew members and more than 50 people on the ground at the crash site were killed.

Engine fuse pins break during flight

The El Al Boeing 747 catastrophe brought to public attention a hazard that air safety officials had been trying to solve for nearly 15 years: the propensity for corroded or cracked engine mounting safety bolts—called fuse pins—to fail during flight. The disaster also revealed that aircraft designers miscalculated how an engine under high thrust will behave if it comes loose in flight. They determined that if an engine were to break loose, it will safely fall down and away from the aircraft. However, the El Al jet's inboard right engine veered to the side, smashing into the outboard right engine.

The Boeing 747's fuse pins are hollow cylinders of cadmium-plated steel 4 inches long and 2.25 inches in diameter. These fuse pins are designed to be sturdy enough to endure enormous stress from multiple directions throughout the flight's trajectory—as the airplane takes off,

Corroded Boeing engine mounting bolts cause both engines from the right wing of a 747 to fall off. The stricken 747 crashes into an apartment building, killing its crew of 6 and more than 50 residents.

After losing its two right engines, the El Al Boeing 747 slammed into a ten-story Amsterdam apartment building. The plane's crew and more than 50 people on the ground were killed.

climbs, banks, bounces, descends, and lands. Four fuse pins connect the engine strut to the wing. The pins slip through lug fasteners on the strut and then through parallel lugs on the wing.

Fuse pins are built to break—under special conditions

Like automobile electrical fuses, these Boeing fuse pins are designed to be fragile enough to break under certain circumstances. For example, designers intend the fuse pins to snap if an engine starts coming apart in flight, so that the damaged engine will sheer cleanly away without damaging the wing. Similarly, if a 747 were forced to make a belly landing, the design specifications call for the engines, upon striking the runway, to sheer away from the wings. This will greatly reduce the chance of the hot engines igniting fuel, which can spill from the wings' storage tanks.

From 1969 through 1980, Boeing mounted its 747 engines with fuse pins honed by machinists to just the proper strength by boring an intricate hourglass shape across the interior surface of the pin. But this original technique turned out to be flawed. Officials discovered that the machining process sometimes left tiny nicks where corrosion could develop. Under stress, the corroded nicks developed into cracks that made the pins dangerously weak. In 1979, therefore, the Federal Aviation Administration (FAA) ordered airlines to periodically inspect these "old-style" fuse pins for corrosion and cracks.

Boeing tried to correct the problem by designing a "new-style" pin featuring a simplified hourglass core supported by inserts wedged permanently into the two open ends of the fuse pins. Factory installation of the new-style pins on new jetliners began in 1980. In 1982 the FAA informed airlines operating older 747s that regular inspections of fuse pins would no longer be required if the old-style fuse pins were replaced with the newer type.

Then reports soon reached Boeing that the new-style pins were also corroding and cracking. In 1985 or 1986 an alert ground inspector spotted an engine drooping on a 747 passenger jet just minutes before it was to take off. The flight was aborted, and inspectors discovered that the droop was caused by a cracked new-style fuse pin. Over the next few years several more cases of weakened new-style pins occurred. In May 1991, therefore, the FAA ordered airlines operating 747s equipped with the new-style pins to perform a one-time check to determine whether the anti-corrosive primer on the inside core of the pins was intact. If corrosion or cracks were discovered, the airlines were instructed to install fresh pins.

Source of fuse pin corrosion is located

The FAA never formally tallied or assessed the cases of corroded or cracked pins uncovered by the inspection. Moreover, some airlines balked at the requirement, because they had to squeeze this time-consuming task into their rigorously tight maintenance routine. Mechanics were not pleased with the requirement either. In order for the inspection to take place, the tightly wedged inserts had to be yanked loose with a special tool to check for corrosion or cracks inside the pin.

As it turned out, however, the inserts themselves were the reason the new-style pins were susceptible to corrosion. Upon being wedged into place in the ends of the pins, the inserts scraped away small amounts of the anticorrosive primer coating the internal bore. Moisture, probably from intense condensation, then seeped into the bore and corroded the scraped area.

On December 29, 1991—eight months after the FAA announced the inspection schedule that would "ensure" the safety of the 747 fleet—a 12-year-old China Airlines 747-200 jet freighter crashed near Taipei, Taiwan, killing all five crew members. The 747's original, old-style fuse pins had recently been replaced with the new-style pins. Even so, a few minutes after takeoff from Taipei, the airplane's right inboard engine ripped loose and slammed into the right outboard engine, knocking it, too, off the wing. Both right-side engines fell into the sea, and the airplane, stuck in a right-hand turn, smashed into a cliff.

New fuse pins suffer same problems as old

For several months, very little information was made public about this Boeing 747 crash in Taiwan. Then, on September 11, 1992, an inspector checking an Argentina Aerolineas 747-200 passenger jet prior to takeoff noticed that the aircraft's right inboard engine was sagging, and mechanics identified a cracked new-style pin causing the droop. A week later Boeing gathered several major 747 operators to meet in Seattle, Washington, to discuss the fuse pin problem. Once again the airlines were directed to perform a one-time inspection of the new-style fuse pins, which was to take place October 8, 1992.

Details of the Crash

On October 4, 1992, however, at 6:22 P.M., the El Al Boeing 747-200 jet freighter, laden with 114 tons of commercial cargo, took off from Amsterdam. Heading southeast over the Dutch city, the airplane was bound for Tel Aviv, with Captain Isaac Fuchs at the controls. At around 6:25 the jet reached 4,000 feet; it continued climbing under high thrust.

A minute later, Captain Fuchs reported to the control tower that a fire warning light for the right inboard, or No. 3, engine was on. An engine separating from the wing could trigger this warning light—and later wreckage analysis confirmed that the No. 3 engine fuse pin failed, setting the accident into motion. At 6:28 Fuchs reported the right outboard, or No. 4, engine out. Later, inspectors would verify that the No. 4 engine operated properly until the No. 3 engine knocked it off the wing.

A large piece of the 747's wreckage is guarded by police in Amsterdam. Investigations showed that corroded engine mount systems allowed the two right engines to drop off the wing, causing the catastrophe.

Disabled 747 crashes into apartment building

Fuchs had no way of knowing from the control panel that the jet had completely lost its two right-side engines. By 6:34, after flying a wide right hand circle over Amsterdam, Fuchs reported a "problem with the flaps," the movable control surfaces on the trailing edge of the wings. Only then did Fuchs declare, "We have a control problem." At 6:35 Fuchs radioed: "Going down, 8162 going down, going down." The jetliner slammed into the ten-story apartment building at 6:35:53 P.M., sundering the building in two. Along with the 6-member El Al crew, more than 50 people at the apartment building site were casualties of the crash.

The El Al jetliner was one of about 200 older-model 747s that had continued to use the old-style pins. Airline records showed that El Al mechanics inspected the pins four months before the accident, per regulations, and reported no problems. Israeli authorities made their determination in January 1993, based on evidence uncovered in the wreckage. They blamed the broken fuse pin on the inboard right engine for allowing the engine to tear off and veer into the outboard engine. El Al officials reported no signs of sabotage and maintained that the crew did everything possible to maintain control of the damaged airplane. An El Al director declared that he would demand payments from Boeing for the victims' families and, given the damage to its reputation, to the airline itself.

Impact

The danger posed by both versions of the Boeing 747 fuse pins was long-recognized. Considering that the El Al Boeing 747 catastrophe came just 11 months after the crash of the China Airlines 747 freighter in Taiwan, airline authorities should have acted more quickly and definitively to address the problem. After the Amsterdam accident, the FAA and Boeing did take more aggressive steps to identify and replace weak fuse pins in the 747 fleet of more than 900 jets. However, it is still unknown if the design problems of the 747 engine mount system have been solved.

Fleet inspection turns up 20 percent defect rate

The day after the crash, Boeing issued the advisory written in connection with the crash in Taiwan, which called for a one-time re-inspection of new-style pins. Boeing exempted the newer model 747-400 from the advisory, however, and gave no recommendations regarding 747s still equipped with old-style pins. A few days later, the FAA widened the call for inspection by including the 747-400 model and requiring more frequent inspections of 747s still using the old-style pins. Inspections of the first 300 jets turned up 499 corroded pins and 14 cracked pins, a 20 percent defect rate. Airlines were permitted to refurbish lightly corroded pins, but heavily corroded and cracked pins were required to be replaced.

Boeing has also continued its development—which was initiated several months before the El Al crash—of a corrosion-proof fuse pin. This has proven to be a difficult design goal. Like other exterior parts, the fuse pins are exposed to intensive condensation as a jet rapidly ascends to and descends from the subarctic levels of cold found in jetliner cruising lanes seven miles above Earth. Other potential sources of corrosive moisture include de-icing agents and various types of liquid cleaners and polishes routinely sprayed on the wings near the engine mounts.

Americans use breakaway engines, which French reject

Boeing uses a 747-type engine mount system, including fuse pins, for all its models. As of late 1992 Boeing was working to address widespread problems with fatigue cracking of 757-model fuse pins. The design philosophy of the French Airbus airplane manufacturer, in contrast, dictates that engines remain mounted to the wings under all conditions, including massive engine failure and emergency belly landings. Breakaway fuse pins are not used on Airbus engine mounts.

Loose engine trajectory is an unsolved problem

Boeing officials declined to publicly discuss the separate phenomenon of a loosened engine, which—contrary to design expectations—veers into the other engine mounted on the same wing. Aeronautical experts theorize that substantial gyroscopic forces can cause a suddenly loose engine to rocket forward in an arc either to the right or the left, depending on the direction of engine fan blade rotation. This new information about the trajectory of loose engines has raised a separate concern: the increasing use of jetliners with only one powerful, wide-diameter engine under each wing. These models include the Airbus A300, A310, A321, A330, and A340, and the Boeing 737, 757, 767, and 777. Could a loose engine on one of these jets strike the fuselage—the central portion of a passenger-carrying airplane—and severely damage it?

Where to Learn More

Acohido, Byron. *Seattle Times* (October 9, 1992; October 17, 1992; October 22, 1992; November 3, 1992; November 4, 1992; November 14, 1992; December 27, 1992; December 28, 1992; December 31, 1992; January 9, 1993; January 16, 1993).

Egozi, Arie. "El Al Crash Centered on 'Engine-pylon Fittings.'" *Flight International* (October 28–November 3, 1992): 5.

Federal Aviation Administration. *Emergency Airworthiness Directive No. T92-24-51.* Washington, DC: Government Printing Office, November 13, 1992.

"Fuse Pins Suspect in El Al Crash." *Aviation Week and Space Technology* (October 12, 1992): 30–33.

Hornblower, Margot. "Death from the Sky." *Time* (October 19, 1992): 50–52.

Lane, Polly. *Seattle Times* (October 13, 1992; October 15, 1992; October 30, 1992; November 13, 1992; January 11, 1993).

Mecham, Michael. "El Al Strut Shows Possible Fatigue." *Aviation Week and Space Technology* (October 26, 1992): 30.

National Transportation Safety Board. *Safety Recommendation Nos. A-92-114–117.* Washington, DC: Government Printing Office, November 3, 1992.

Nelan, Bruce W. "Are 747s Safe to Fly?" *Time* (October 19, 1992): 52.

Sharn, Lori. "Faulty Pins Found on U.S. 747s." *USA Today* (October 20, 1992): A6.

West, Karen. "Boeing Suggests Checkups: Airlines Advised to Inspect Steel Pins." *Seattle Post-Intelligencer* (October 6, 1992): A6.

Apollo 1 catches fire

Cape Canaveral, Florida
January 27, 1967

Background

Three astronauts were killed during a routine ground test of the Apollo command module on January 27, 1967, at the National Aeronautics and Space Administration (NASA) base in Cape Canaveral, Florida. Although the exact cause of the fire remains undetermined, it probably began with an electrical arc caused by poor wiring design and installation. Once some combustible materials ignited, the fire was fed by the pure oxygen under pressure in the module. The astronauts had no firefighting equipment and were unable to open the six-bolt escape hatch before they died of asphyxiation.

Project Apollo's mission is to land men on the Moon

The United States was racing against the Soviet Union to land men on the Moon. After the successful two-man Gemini program, the three-man Apollo capsules would make a Moon landing possible. AS-204 was to be the first manned Apollo mission. Its crew was named in August 1966: Two space veterans, Virgil I. "Gus" Grissom and Edward H. White, and one new astronaut, Roger B. Chaffee, who had not yet flown in space. The crew would make a December 1966 "shakedown" flight of up to two weeks in Earth's orbit. This flight would test the Apollo's new command and service module (CSM). These were built by North American Aviation and had not been deployed on a manned mission. Following a series of mechanical problems, however, the flight date was postponed until February 1967.

Prior to actual flight, the astronauts were to complete a four-phase ground test of the CSM's systems. One phase, a "plugs out" test to see if

Three astronauts die during a ground test when their command module suffers a flash fire. Evidently due to a wiring malfunction, the fire exposes a need for higher design, manufacturing, and safety standards at NASA.

Apollo 1 crew members Virgil I. "Gus" Grissom, Edward H. White, and Roger B. Chaffee answer questions at a 1966 press conference in California. All three perished during a routine ground test of the Apollo command module.

the spacecraft could run on its internal power system, required the astronauts to enter the command module. The CMS was perched on a Saturn IB launch vehicle and emptied of fuel. The astronauts would be sealed in the module, it would be pressurized, and the test would begin.

Manned cabin to be oxygenated at 16 psi

The command module was designed to function in space. In that setting it would be subject to greater internal than external pressure. To simulate this condition as closely as possible, NASA's engineers planned to pressurize the cabin with pure oxygen at 16 psi (pounds per square inch), rather than the normal 5 psi the CSM would usually be subject to in space. Although pure oxygen itself will not ignite, it can, when pressurized, rapidly feed an existing fire. No one considered a cabin fire likely, and there was no fire extinguisher in the capsule. Since the launch vehicle was not fueled, fire crews would be on standby rather than maximum alert. And no one considered excessive the 90 seconds it took to open the new six-bolt escape hatch. The old quick-release hatches used on the Mercury and Gemini capsules were considered accident-prone and dangerous.

NASA's confidence about safety precautions stemmed from the fact that, in six years of manned space flights, no astronaut was ever killed in the course of a mission. The "plugs out" test, conducted with an unfueled launch vehicle, seemed both routine and low-risk.

Details of the Fire

On January 27, 1967, Gus Grissom, Edward White, and Roger Chaffee commenced the "plugs out" test. By 1:00 P.M. on that Friday afternoon, the three-man crew had crawled through the open hatch and assumed their flight positions. After two hours of tests conducted with the hatch open, the capsule was sealed and the cabin pressurized with pure oxygen to 16.2 psi. The crew then went through a practice countdown and ran simulation tests for more than three hours, regularly interrupted by minor problems.

Telemetry reports major electrical short

Shortly after 6:00 P.M., 15 minutes before liftoff was to be simulated, the spacecraft switched to internal power. NASA engineers called a hold to check on some problem, and the astronauts waited yet again. The astronauts had been in their couches some five and a half hours; it had been a long day. Suddenly, at 6:31 P.M., telemetric data from the spacecraft indicated that a major short had occurred somewhere in the nearly 12 miles of electrical wiring packed into the command module. Less than 10 seconds later, Roger Chaffee made an almost casual report: "Fire, I smell fire."

Capsule explodes

The spacecraft did not have internal cameras, but a camera outside was focused on its porthole. At the first report of fire, the camera operator

The flash fire that spread through the *Apollo 1* spacecraft—killing all three occupants—was probably caused by an electrical arc near the environmental control equipment under Grissom's couch.

saw nothing more than a sudden bright glow. Then he saw flames flickering across the porthole and Edward White's hands reaching above his head to get at the bolts securing the hatch. The operator saw a lot of movement, and then another pair of arms struggling with the hatch. Soon dark smoke completely obscured the scene. The last sound from the astronauts was a frantic cry from Roger Chaffee: "We've got a bad fire—let's get out . . . we're burning up!" Seconds after this last transmission, the tremendous pressure inside the cabin split the capsule open, and a blaze of flame gushed out. Only 18 seconds elapsed between the first call of fire and the explosion.

While there was help was close at hand, the control personnel in the white room (a dust-free room in the service tower just outside the command module's hatch) were momentarily held back by the explosion. Due to the thick smoke, it then took five men working in shifts 5½ minutes to remove three separate hatches: the boost protective cover, which shielded the command module during launch; the ablative hatch; and the inner hatch. After the smoke thinned, they found the three men dead inside the capsule. Chaffee was still strapped onto his couch, and White and Grissom were lying close together below the hatch. White's handprint was outlined in ash on the hatch.

Two astronauts were welded to capsule floor

Official autopsies later identified the cause of the deaths as asphyxiation. These post-mortem reports observed, however, that although each astronaut had suffered serious burns, they were not fatal burns. The heat during those few terrible seconds had been so intense—the holes burned in aluminum tubing indicate temperatures of at least 760°C—that the astronauts' suits had melted and fused with the molten nylon and Velcro inside the capsule, forming a synthetic liquid that solidified as it cooled. Doctors arrived 14 minutes after the first alarm of fire, but it took them seven hours to remove the bodies. Those of White and Grissom had been welded to the capsule floor.

Impact

NASA's multi-billion dollar effort to put a man on the Moon virtually stood still as a special Board of Inquiry sought explanations for the accident. On February 22, 1967, almost a month after the disaster, a seven-man review board issued an interim report stating that although no definite cause of the fire could be established, the most likely origin was an electrical malfunction. A 14-volume report of some 3,000 pages came out in early April 1967. This report specified that the fire was probably caused by an electrical arc that occurred near the environmental control equipment under Grissom's couch.

Report criticizes manufacturing and lack of safety

The full report was highly critical of the conditions that enabled the fatal accident. Analysis of some of the wiring unaffected by the fire revealed "numerous examples of poor installation, design, and workmanship." Moreover, the capsule was loaded with highly combustible materials such as Velcro and the nylon netting used to prevent loose objects from floating around in the zero gravity of space. The report concluded that disaster may have been inevitable, given the substandard manufacturing procedures and the lack of safety measures.

The negative critique of NASA was not lost on Congress, and a House space subcommittee opened hearings on April 7, 1967, to assess the space program and the Apollo accident. Most observers decided that NASA had been pushing the Apollo program too hard and too fast without thoroughly testing all its systems. The Apollo spacecraft was still an unproven and evolving craft: it underwent 623 changes between August 1966—when it was delivered by North American Aviation to NASA, and January 1967—when the fatal accident occurred. The prevailing conclusion was that it would be ill-advised to allow the space program to continue on its timetable. An exhaustive review of the entire spacecraft was necessary, however long it took.

NASA redesigns command module

NASA was obligated to step back and reassess the full scope of its systems and procedures, particularly in terms of safety standards. The 18-month hiatus from its rivalry with the Soviet space program enabled NASA to produce a completely redesigned Apollo command module with some 1,500 modifications, resulting in a considerably more secure and fireproof vehicle.

First, NASA installed high-quality wiring. Flameproof coatings were applied over all wire connections, plastic switches were replaced by metal ones, and soldering became more meticulous. Almost all flammable materials inside the module were removed. A new, fire-resistant material known as Beta cloth was developed for spacesuits. Instead of igniting at 500°C, as the old Nomex suits had, the Beta cloth suits could withstand temperatures of more than 800°C.

Considerable debate over the use of pure oxygen resulted in a compromise in favor of safety. When in space, the crew would breathe pure oxygen at 5 psi. For ground testing and launching, however, the cabin would be filled with a mixture of oxygen and nitrogen at sea-level pressure (14.7 psi). To prevent this nitrogen from causing "bends" (nitrogen narcosis, or poisoning), the astronauts would breathe only through their spacesuits, which contained pure oxygen, until the cabin had been purged of nitrogen.

Capsule door would not have opened

One irony of the Apollo accident was that for their last task that day, Grissom, White, and Chaffee were to test the new six-bolt escape hatch. This system replaced the quick-release, explosively charged hatch that had been used on both Mercury and Gemini spacecraft. In 1961 the quick-release hatch had been blamed for prematurely blowing and almost sinking Grissom and his Mercury *Liberty Bell 7* when he splashed down in the Atlantic. After the disaster, Apollo returned to an improved quick-escape system. The modified hatch took only 12 seconds to release and opened outward, so internal pressure would have no effect on its functioning. The full report on the Apollo accident revealed that even if the astronauts had managed to undo the six-bolt hatch, their efforts would have been futile. When the internal pressure (which reached 16 psi) exceeded the external pressure (14.7 psi) by just 0.25 psi, the hatch, which swung inward, became impossible to open.

If good can be said to come of such a disaster, it was that the American government and industry coalition became conscious of the need to raise design, workmanship, and safety standards. The fire both enabled and forced NASA to step back from its politically controlled timetables. The high-performance spacecraft that ultimately put Americans on the Moon would be more methodically built.

Where to Learn More

Bond, Peter. *Heroes in Space: From Gagarin to Challenger.* London: Basil Blackwell, 1987.

"Electrical Malfunction Termed a Likely Source of Apollo Fire." *New York Times* (February 26, 1967): 39.

McAleer, Neil. *The OMNI Space Almanac: A Complete Guide to the Space Age.* New York: World Almanac, 1987.

Murray, Charles, and Catherine Bly Cox. *Apollo: The Race to the Moon.* New York: Simon and Schuster, 1989.

Wheeler, Keith. "Disaster: The Harsh Schoolmaster." *Life* (January 26, 1968): 56–59.

Wilford, John Noble. "Apollo Fire Review Boards Finds 'Many Deficiencies'; Calls for Safety Moves." *New York Times* (April 10, 1967): 1.

Soyuz 1 crashes

Orenburg, Russia
April 24, 1967

First fatal crash of Soviet manned space program is blamed on tangled parachutes; additional failures probably involved flight and reentry control, which sent the U.S.S.R. program into decline at a crucial phase of the "space race."

Background

Veteran cosmonaut Vladimir M. Komarov was killed when reentry of his *Soyuz 1* spacecraft into Earth's atmosphere could not be controlled. This was the first fatality in the Soviet space program. The flight of this new class of Soviet spacecraft was apparently troubled from the onset, although Soviet officials maintained that the mission proceeded as planned until after reentry. Sudden tangling of the spacecraft's landing parachutes caused its crash.

Western experts attempted to reconstruct fully what happened, and they concluded that solar panel and antenna malfunctions probably led to a major control failure during orbit. They also surmised that "tumbling" of the spacecraft during its forced reentry could have made the parachute foul-up and crash landing inevitable. Most observers believe that political considerations played a major role in fatally rushing the *Soyuz 1* launch date.

With *Soyuz,* Soviet cosmonauts would really be pilots

The Soviets launched *Soyuz 1* on April 23, 1967, after a hiatus of more than two years from manned space flight. The Soviets had taken this time to develop the new Soyuz class of manned spacecraft. Soviet cosmonauts had flown in the older Vostok and Voskhod spacecraft, but these spacecraft had automatic controls and were essentially operated from the ground. Soyuz, meaning "union" in Russian, would be used for sophisticated rendezvous and docking operations and rely on cosmonaut control of the navigation systems. For the first time, Soviet cosmonauts truly would function as pilots.

Veteran cosmonaut Vladimir Komarov was killed when reentry of his *Soyuz 1* spacecraft into the Earth's atmosphere could not be controlled.

During the two-year hiatus in Soviet manned flights, the United States successfully executed ten consecutive two-man launches in the Gemini series. The Americans were pulling significantly ahead in the race to put a man on the Moon until the fatal *Apollo 1* fire on January 27, 1967. The American disaster put the Soviet Union (now the former Soviet Union) and their new Soyuz spacecraft back into the competition. The

Even if he had remained conscious, Komarov probably would have been unable to stop the spacecraft's rotation before the parachute deployed four miles above Earth. The spinning would have caused the shroud lines to become tangled around the spacecraft, preventing the parachute from unfolding or slowing down the spacecraft's rate of descent.

Soviets entrusted the first Soyuz mission to veteran cosmonaut Vladimir Komarov; he would become the first Soviet to make a second trip into space.

Soyuz would be powered by solar cells

The *Soyuz 1* spacecraft was 7.9 meters long, consisting of an orbital module, a descent module, and an equipment module. Weighing nearly 400 kilograms, it would be launched from the Baikonur space center by the A-2, a standard Soviet launch vehicle. After achieving orbit, the launch vehicle would deploy a set of solar panels, which would unfold like wings. The 14 square meters of cells on the panels would power the spacecraft's systems.

Soyuz crews would spend their mission in the orbital module, but they would reenter Earth's atmosphere in a descent module after jettisoning both the orbital and the equipment modules. To a degree, the American Apollo spacecraft relied on a similar system. In the American system, however, astronauts would "splashdown" into the sea and be retrieved by helicopters operating from an aircraft carrier. The Soviets opted to land their spacecraft, equipping it with two sets of parachutes and a special retro-rocket system to cushion the impact.

The *Soyuz 1* was a technologically groundbreaking mission, and the Soviets were uncharacteristically open about their plans. In conferences granted to the Western press, Soviet officials boasted that the mission would involve two separate manned spacecraft, a rendezvous and docking, and perhaps a crew transfer.

Details of the Crash

Soyuz 1 was launched April 23, 1967, at 3:35 A.M. Moscow time, the first Soviet manned spacecraft to be launched at night. Several hours after *Soyuz 1* entered into its nearly circular orbit 200 kilometers above Earth, Soviet television and radio suddenly became silent about its progress. Even so, the Western world had little indication of trouble with the flight until a day later, when the scheduled launch of the second Soyuz craft—which the orbiting spacecraft presumably was to link up with—did not occur. Subsequently, at 6:00 A.M. on April 24, Komarov was ordered to return to Earth. At this point, Soviet officials announced that their modest one-day test flight had proceeded according to expectations.

Cosmonaut Komarov reported dead in "freak accident"

After 12 more hours of ominous silence, Moscow announced that a freak accident had caused the spacecraft's parachute lines to tangle just prior to landing, causing a fatal crash. Komarov was the first cosmonaut to be killed during a space mission, and he was buried with full honors.

Although the *Soyuz 1* crash was a global news event, the Soviets released no further details about it. Officials maintained that nothing untoward had happened during the mission, and that the freak tangling of the parachute lines was alone responsible for the disaster.

Western experts reconstruct details about flight of *Soyuz 1*

Western experts have analyzed the available information about the accident to determine what happened to *Soyuz 1*. Most agree that the Soviets did not launch the second Soyuz because of a problem with *Soyuz 1*, already in orbit. Additionally, there is evidence that one of the two solar panels of the original spacecraft did not deploy properly when the craft reached orbit. This meant that the panel's thermal radiator would be covered, reducing the spacecraft's electrical power supply. Most experts also believe that the spacecraft's television transmitter must have failed and that other antennas did not deploy properly.

The most serious problem, however, was an evident malfunction of the spacecraft's altitude control and guidance systems. This malfunction has been attributed to an overheating caused by the still-folded solar array, which might have prevented dissipation of the heat generated by the spacecraft's electronic systems.

Spacecraft tumbled end-over-end in its orbit

Soyuz 1 also came down far short of its planned landing site—600 miles off course, near the city of Orenburg on the steppes north of the Caspian Sea. This fact supports the control failure hypothesis. Studies have shown that the 600 mile undershoot could have been caused by the spacecraft's spinning during reentry. If a spacecraft's guidance systems fail, moreover, a pilot might deliberately trigger such a spin or roll in a desperate maneuver to stabilize the craft. U.S. tracking facilities documented that *Soyuz 1* was tumbling or slowly rotating end-over-end during its fifteenth orbit.

By the seventeenth orbit, Komarov had been given permission to reenter. Evidently, however, he could not steady the spacecraft sufficiently to fire the reentry rockets accurately. Komarov apparently initiated the spin or controlled roll maneuver on the next orbit, and then fired the reentry rockets with the hope that the roll would balance all the aerodynamic forces set into play. The danger of this maneuver, however, is that the spinning descent that resulted could subject Komarov to as many as 10 g's (10 times the force of gravity), double the normal reentry g's of 4 to 5. Thus, Komarov might have lost consciousness during reentry.

Without chutes, *Soyuz 1* probably hit Earth at 300–500 mph

Even if he did not black out, Komarov probably was unable to stop the rotating of *Soyuz 1* before the parachutes deployed four miles above Earth. The spinning would have caused the shroud lines to become tangled around the spacecraft and prevented the parachute from unfolding or slowing down the spacecraft's rate of descent. The same factor would have rendered the emergency parachute useless. Whatever the case, there is little doubt that Komarov and the spacecraft hit the ground at a tremendous speed—between 300 and 500 mph (miles per hour). No details about the wreckage have been released, but such a high-impact crash would have devastated the spacecraft.

The reluctance of Soviet officials to release details about what went wrong with the mission allowed all sorts of stories to flourish. One of these was the account of a U.S. National Security Agency technician, who years later stated that he had overheard *Soyuz 1* spacecraft-to-ground communications from his listening post in Turkey. He claimed that Komarov, fully aware of the dangers that reentry would involve, protested to Soviet ground control: "You've got to do something. I don't want to die." This technician also reported that Komarov was allowed to speak with his wife one last time. A separate account from Italy has Komarov screaming at the beginning of the fatal reentry maneuver: "You are guiding me wrongly.... Can't you understand?"

Impact

Years after the crash, a Soviet engineer who immigrated to the United States discussed the *Soyuz 1* mission in his memoirs. He indicated that the Soyuz spacecraft was launched before all the design bugs had been eliminated. The author further charged that repeated failures had occurred during preflight tests, especially in the crucial thermal control and parachute systems, and that everyone involved knew of these malfunctions.

The author reported that the deputy chief designer (named Mishin) in charge of the space mission refused to sign the preflight papers endorsing the launch. Over the objections of the technical staff, however, the political arm of the Soviet government ordered the *Soyuz 1* mission to proceed. Soviet politicians feared that they would otherwise fall even further behind in their "space race" with the United States.

After tragic loss of Komarov, Soviet space program stalls

The subsequent loss of *Soyuz 1* signified more than the failure of an individual mission. The Soviet government and people were stunned and humiliated by their first space disaster. Given the international embarrassment surrounding the accident and the technological problems that obviously remained to be solved, it is not surprising that the Soviet manned space program entered a period of decline. The next test flight of a Soyuz spacecraft would not take place until 18 months after the disaster. With this test flight, the Soviets were just beginning to accomplish the manned orbital maneuvers that the Americans had achieved long before in their highly successful Gemini program. Twenty-seven months after Vladimir Komarov was killed, the United States became the first country to land men on the Moon and return them safely to Earth.

Where to Learn More

Bond, Peter. *Heroes in Space: From Gagarin to "Challenger."* London: Basil Blackwell, 1987.

Clark, Evert. "Soyuz Problems Discerned in U. S." *New York Times* (April 28, 1967): 20.

Newkirk, Dennis. *Almanac of Soviet Manned Space Flight.* Houston, TX: Gulf Publishing, 1990.

Oberg, James E. *Red Star in Orbit.* New York: Random House, 1981.

———. *Uncovering Soviet Disasters: Exploring the Limits of Glasnost.* New York: Random House, 1988.

———. "Vladimir Komarov: First Space Casualty." *Space World* (June 1974): 4–15.

"U.S. Aides Believe *Soyuz 1* Was Out of Control." *New York Times* (April 26, 1967): 3.

Apollo 13 oxygen tank explodes

April 13, 1970

An exploding oxygen tank in the service module jeopardizes the crew's lives as they are forced to use the lunar module to get back to Earth. Congressional budget cuts curtail later Apollo missions.

Background

On April 13, 1970, two days after the three-man crew on *Apollo 13* left Earth for the Moon, an oxygen tank explosion disabled their spacecraft. The astronauts had to abandon their planned lunar landings and get back to Earth under very precarious conditions. With their command module steadily losing power, the crew transferred into the cramped lunar module and carefully rationed power, oxygen, and water supplies for the four-day trajectory around the Moon and back to Earth. Taking various measures to survive the trip back, the astronauts used the lunar module in ways never before attempted. Just prior to reentry, they transferred back into the command module, powered it for a last time, separated from their lunar module "lifeboat," and returned to Earth.

Apollo 13 plans ambitious mission with two lunar walks

Apollo 11 and *Apollo 12* were highly successful manned lunar landing flights. *Apollo 13*—with two moon-walks scheduled—was an even more ambitious mission. The spacecraft was made up of paired command and service modules (CSM); the command module housed the crew, the service module contained the engine and supply systems. The independent lunar module was attached to the command and service modules. This third module was designed to be used for the Moon landings while the CSM continued in a lunar orbit. The lunar module would then blast off from Moon, relink to the command and service modules, and eventually be discarded.

The crew was commanded by James A. Lovell, a veteran of two Gem-

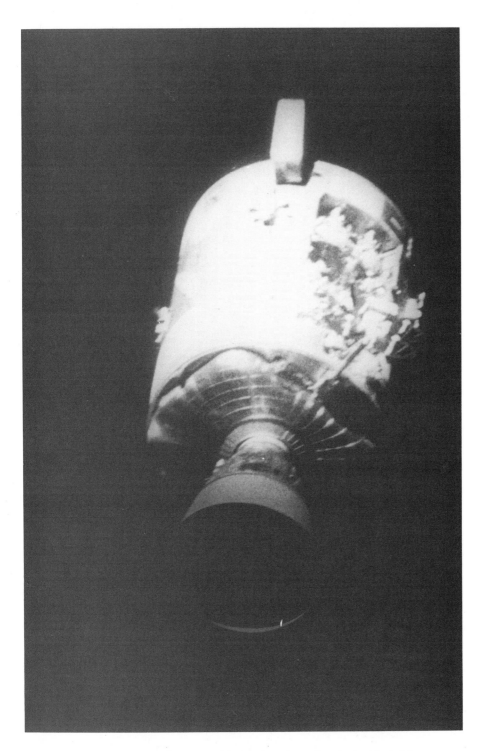

Taken by the Lunar Module/Command Module four days after the incident, this photograph shows the severe damage incurred after the explosion of one of the craft's oxygen tanks. The crew members inside were forced to evacuate and use the lunar module as a "lifeboat."

ini missions and the first circumlunar flight, *Apollo 8*. He was joined by two rookie astronauts: Fred W. Haise, the lunar module pilot, and John L. Swigert, the command module pilot. Swigert was a last-minute replacement for Thomas K. Mattingly, who had been exposed to German measles.

Bad luck will plague thirteenth voyage

The *Apollo 13* flight both tested and confirmed the bad luck connotations of the number 13. The thirteenth Apollo voyage was launched at the thirteenth minute of the thirteenth hour of the day, Houston time. On April 13, two days after its launch, the crisis occurred.

Details of the Crisis

The crisis that handicapped *Apollo 13* took place in the initial part of the mission, approximately 56 hours and 205,000 miles into the flight. The astronauts heard a bang, after which the pressure dropped in the service module's oxygen tanks and then in the command module. John Swigert matter-of-factly reported the malfunction in the service module oxygen tanks: "Houston, we've got a problem." No one in the spacecraft or at the NASA (National Aeronautics and Space Administration) base in Houston understood why the oxygen pressure gauge for tank 2 soon read "zero" and the needle to the tank 1 gauge was dropping.

Losing oxygen supplies will compromise mission

Thirteen minutes after the bang, the *Apollo 13* crew noticed a white, wispy cloud beginning to surround the service module. They correctly guessed that it was their oxygen. Necessary not only for breathing, oxygen enabled the fuel cells in the service module to produce electrical power and water. Its uncontrolled venting explained not only the pressure loss in the oxygen tanks but also the fact that they kept veering off course. Debris from the explosion would gradually encircle the spacecraft and become an additional hazard.

The controllers in Houston struggled to interpret their telemetry. The data was telling them that two redundant Apollo systems—fuel cells and oxygen tanks—were failing. No one yet realized how serious the problem actually was. Instead they attributed it to some sort of system breakdown or failure.

Crew will have to use lunar module to save their lives

Nevertheless, both Houston and the *Apollo 13* crew comprehended that the central command and service modules were becoming inoperative. About 75 minutes after the venting, the only possible course of action became clear: The command and service modules had to be shut down. The fuel cells had only about 15 minutes of power left—they had to save it to reenter Earth's atmosphere. The crew thus deactivated the two modules and powered the lunar module for use as a survival craft. While this "lifeboat" option had been anticipated by the builders of the Apollo spacecraft, no one had yet tested this strategy.

From the time the crew transferred into the small lunar module until they got back to Earth, the NASA controllers in Houston had to improvise nearly every procedure. The lunar module was designed for the use of two men for 33 to 35 hours; this emergency would require it to accommodate three men for up to 100 hours. Houston also had to calculate how to use the lunar module's thrust engines to propel a complex spacecraft configuration.

Astronauts are subjected to near-freezing temps, CO_2 buildup

All power, oxygen, and water supplies on the lunar module had to be strictly rationed. Power consumption was reduced to one-fifth of the normal level. Conserving power included not heating the spacecraft—its temperature dropped to a near-freezing 38°F. As it got colder inside the modules, the astronauts' breath formed frost on the inside of the windows. The cold became almost unbearable, and Fred Haise developed a kidney infection that plagued him for the rest of the trip.

The buildup of CO_2 (carbon dioxide) became a serious threat. The lithium hydroxide canisters in the lunar module could not remove all the CO_2 produced by three men. Canisters from the command module were useless because of their different shape, so the crew devised a makeshift system to neutralize the CO_2.

Other hardships affected them. The astronauts were restricted to six ounces of water each per day. They wouldn't starve, but they couldn't reconstitute their food packs because there was no hot water. In addition, they couldn't eject any waste into space for fear that venting might affect the spacecraft's delicate trajectory back to earth.

Navigating the crippled spacecraft required its own set of improvisations. When the craft reached the far side of the Moon, the lunar module's engines had to be fired very precisely to free the spacecraft from the

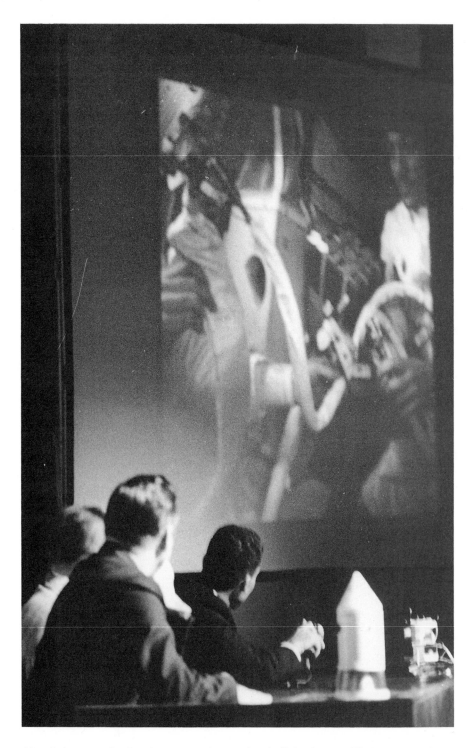

After their return, *Apollo 13* crew members review in-flight photos of Swigert preparing one of the supplemental lithium hydroxide canisters. When canisters on board began to fail, the crew had to devise a makeshift neutralizing system to eliminate the dangerous CO_2 buildup.

Moon's gravitational pull and send it back toward Earth. After this maneuver, three other engine firings were crucial to speed the craft's return to Earth and to correct its course. Executing these mid-course corrections was delicate and tense, because debris from the original explosion now encircled the spacecraft like a cloud and gleamed in the Sun. These glittering fragments made it impossible for the astronauts to see familiar guide stars for navigating and calculating proper alignment. NASA personnel in Houston, however, were able to conduct simulations and devise new strategies to navigate the craft.

Crew suffers oxygen, water, and sleep deprivation

With each day that *Apollo 13* came closer to Earth, NASA controllers became more concerned about the physical state of the crew. In addition to the cold and the shortage of oxygen and water, there was no place to sit or recline in the lunar module, which had only standing room for two men. The third crew member was forced to spend periods of time trying to sleep in the even colder command module. Houston later learned that none of the crew slept more than 12 hours in the three and a half days it took to return to Earth. NASA worried especially about how the fatigue, cold, and dehydration would affect the crew's efficiency, since they would soon have to copy an emergency list of procedures and commands read to them from Houston. This task alone would take two hours.

The astronauts used these commands—which specified hundreds of switch, valve, and dial positions—to prepare for reentry and splashdown. Among their preparations was powering the command module. Its heat-shielded capsule would protect them during the fiery reentry into Earth's atmosphere. If the crew could not activate this module, they would be left stranded in space. When the power-up worked, the astronauts began to feel warm for the first time in days.

Five hours before splashdown on April 17, the lunar module's thrust engines were fired a final time to initiate the reentry. A half hour later the crew used explosive charges to jettison the damaged service module. As it drifted away from them, the astronauts documented the damage on photographs. Swigert exclaimed, "There's one whole side of that spacecraft missing!" The astronauts left the lunar module for the last time. From inside the command module, 90 minutes before reentry, they cast off their lunar module lifeboat. Discounting the deprivations they had endured there, Lovell reflected, "She was a good ship."

Splashdown and recovery operations are accurate and fast

The command module soon reentered Earth's atmosphere. As it streaked through the skies at some 25,000 mph, the heat shield (which ground controllers feared had been damaged in the explosion) withstood temperatures of 4,000°F. After an unusually long period out of contact, the capsule parachuted down within full view of the carrier U.S.S. *Iwo Jima*. *Apollo 13* had the most accurate splashdown and the fastest crew-recovery operation in the history of the Apollo program. After they were rescued from their capsule, the exhausted astronauts could barely stand. Lovell had lost 14 pounds, Haise had a severe kidney infection, and all three men were badly dehydrated. But despite the odds, they had survived.

Impact

As one NASA official soberly remarked, *Apollo 13* reminded everyone "that flying to the Moon is not just a bus ride."

The investigators released their findings in June 1970. The crisis began in the service module's fuel cells, when a heater in an oxygen tank exploded. Both oxygen tanks in the service module burst. Also damaged was the module's main engine and all of its fuel cells.

The cause of the explosion was found to be defective thermostat switches. Several years before the flight of *Apollo 13,* the switches were apparently damaged when the oxygen tank was dropped. In subsequent preflight tests, ground crews turned the heaters and fans on when the tank would not empty. The resulting heat built up, welded the switches, and burned the insulation off the wiring. When the *Apollo 13* crew turned on the stirring fan in the oxygen tank, the conditions were right for an explosion.

In light of the investigation, additional fuel cells and oxygen tanks were designed into the service module, which had been built by North American Rockwell. The lunar module, manufactured by the Grumman Corporation, also underwent design modifications, making it even better equipped to serve as an emergency "lifeboat." NASA also corrected quality control problems in the construction and prelaunch testing of equipment.

Travel to the Moon "is not just a bus ride"

The fact that the *Apollo 13* crisis came to safe resolution showed that thousands of NASA and space industry personnel could cooperate and solve an unprecedented technical dilemma—while three lives hung in the balance. Nonetheless the accident made many observers question the need for repeated manned lunar landings. *Apollo 13*'s oxygen tank rupture set NASA's timetable back nearly ten months, but Congress crippled the program further. Due to congressional budget cuts, *Apollo 18, 19,* and *20* were canceled.

Where to Learn More

"Apollo 13": "Houston, We've Got a Problem." Washington, DC: NASA, 1970.

"Apollo 13: Three Who Came Back." *Newsweek* (April 27, 1970): 21–27.

Cortright, Edgar M. *Apollo Expeditions to the Moon.* Washington, DC: NASA, 1975.

Murray, Charles, and Catherine Bly Cox. *Apollo: Race to the Moon.* New York: Simon & Schuster, 1989.

Strickland, Zack. "Pre-launch Electrical Overload Led to *Apollo 13* Tank Rupture." *Aviation Week and Space Technology* (June 8, 1970): 18–19.

"What Really Happened to *Apollo 13.*" *Space World* (October 1970): 20–37.

Skylab's meteoroid shield fails

Cape Canaveral, Florida
May 14, 1973

Aerodynamic loads during launch of the NASA orbiting space laboratory destroy the meteoroid shield and a solar array.

Background

The unmanned Skylab Workshop was launched on May 14, 1973, from the Kennedy Space Center in Cape Canaveral, Florida. Approximately 60 seconds later the workshop's meteoroid shield prematurely deployed. Although the launch seemed successful in every other way, the mistaken deployment of the meteoroid shield meant that within two hours of launch, Skylab was practically unusable. Nevertheless, ground personnel improvised a workable rescue that saved the Skylab program from becoming a complete loss.

NASA begins new Skylab Program

The Skylab Program was a logical continuation of manned space exploration that began with the Mercury, Gemini, and Apollo programs. Skylab was granted $2.5 billion to create an orbiting laboratory and home-in-space. American astronauts would be able to live and work in a "shirt-sleeve" environment.

Skylab began in 1965 when the National Aeronautics and Space Administration (NASA) proposed a "spent-stage experiment." Astronauts would enter the empty upper-stage section of a Saturn rocket and perform experiments in weightlessness. By 1970 the project had developed into an orbital workshop that would contain four modules, two of which were for laboratory experiments and living space.

Skylab featured four major components:

• A multiple docking adapter for docking the manned spacecraft

A U.S. Customs official in San Francisco examines the largest piece of Skylab. The chunk of debris was shipped to the United States from Australia after the craft fell through Earth's atmosphere into the Indian Ocean (July 1979).

- An airlock module for providing astronaut access to the vacuum of space to perform extravehicular activity
- A solar observatory (Apollo telescope mount) for conducting solar observations and providing power
- The orbital workshop to serve as crew quarters and general work area

Skylab would be sent into Earth orbit in an unmanned launch atop a two-stage version of the reliable Apollo *Saturn V* launch vehicle. This same rocket would later launch the Apollo astronauts to the moon.

Skylab to be largest satellite ever launched

In terms of previous projects, the space station was impressive.

Including its attached command module, which would bring the crews to and from Earth, Skylab was almost 120 feet (36.6 meters) in length—the largest satellite yet launched by the United States. About the size of an average three-bedroom house, it weighed almost 100 tons—on Earth. The orbital workshop area, which included living and workshop quarters, was the station's largest section: 48 feet (14.6 meters) long and 21.6 feet (6.6 meters) in diameter. In Earth orbit, Skylab would be powered by two large arrays of solar cells extending out from the space station like wings.

Details of the Failure

Skylab Workshop blasted off the launchpad in the two-stage *Saturn V* on May 14, 1973, precisely on schedule. As the great rocket began its climb into the atmosphere, all systems were operating normally.

Meteoroid shield accidentally deploys during launch

Then a launch monitor reported, "Flight, we have some indication that the micrometeoroid shield has partially deployed." Skylab's shield was not supposed to deploy until 96 minutes into the flight, yet the computer said otherwise. Premature unlatching of one of the two solar arrays was also reported.

Ten minutes after launch, Skylab separated from the launch vehicle—right on schedule—and 8 seconds later, Skylab began its nearly circular orbit around Earth. After a series of normal adjustments, Skylab was in its desirable "solar inertial" orientation—that is, the centerline of the solar observatory was pointing toward the sun.

Skylab is too hot to inhabit

Meanwhile, ground engineers were monitoring the data from the orbiting Skylab, and their readings were all wrong. The heat on the exterior was 200 degrees above its designed temperature, leading them to suspect that the meteoroid shield had been lost. In addition, motion sensors and loss of temperature and voltage measurements pointed to the loss of the solar array system.

When the planned deployments of the shield and solar arrays failed to occur 41 minutes into the launch, NASA personnel knew they had lost these two vital components. Skylab was in orbit, but it was overheated, underpowered, and uninhabitable.

The meteoroid shield peeled away like a label from a can

When deployed, the meteoroid shield would cover Skylab's cylindrical exterior to protect it from the heat of direct sunlight and from high-speed meteoroid penetration. Unfurled, the shield was a cylindrical structure 270 inches in diameter and 265 inches long. It weighed 1,200 pounds. Passive protection from the sun's heat came from its black and white paint patterns. Protection from meteors came from its thickness—0.025-inch, slightly less than the thickness of a credit card—and from the gap it created between it and Skylab—it stood out 5 inches from the space station's exterior walls. After being deployed in orbit, the shield required very little strength to serve its primary functions. However, since the meteoroid shield was a very limber cover, it would have to be wound very tightly to survive the launch.

Although the meteoroid shield was simple in concept, it was a very complex device. Straps, hinges, and torsion rods held it in place and deployed it in orbit. A special spring-loaded fold-out panel increased its perimeter an extra 30 inches after initial deployment. The torsion rods provided a twisting force, which would rotate the shield to its extended position.

But a different chain of events played out. When the rocket reached 28,000 feet, it was subjected to the maximum aerodynamic forces (or loads) from Earth's atmosphere. At 60.12 seconds into the launch, aerodynamic loads created internal pressures higher than the wrapped shield was designed to bear. Not all the seals were in place on the shield, and those that were present turned out to be deficient. Consequently, as the launch vehicle tore through the atmosphere into space, the thin meteoroid shield was forced into the supersonic airstream, and the hinges that held it in place failed as if they were unzipped. The shield then gave way like a label peeled from a can, tearing one solar array completely off and jamming the other. Skylab arrived in orbit with no protection from the heat of direct sunlight and no solar panels to generate electricity.

Skylab turns into 400°F oven, which delays manned rendezvous

With no meteoroid shield, the sun shone directly onto the workshop's external gold foil coating. Absorbing most of the heat, Skylab could easily overheat to 400°F or more. Besides making the space station uninhabitable, such temperatures would damage on-board equipment. To prevent further overheating, NASA personnel adjusted Skylab's attitude to the sun to 45 degrees. Interior temperatures then stabilized at 130°F. And despite Skylab's profound disability, there was a little electricity available

The shield gave way like a label peeled from a can, tearing off one solar array completely and jamming the other. As Skylab entered orbit, it found itself with no protection from the heat of direct sunlight and no solar panels to generate electricity.

aboard to operate essential equipment. The power came from the Apollo telescope mount solar arrays, which were providing about a third of their full capability.

Original schedules called for the first manned launch to occur the next day. With Skylab crippled, NASA immediately postponed the manned launch for 10 days, then changed the hiatus to 11 days. Eleven days to design, develop, manufacture, and conduct qualification tests—including zero-gravity simulation in the neutral buoyancy facility—was not a lot of time, and NASA employees and contractors worked around the clock to meet the deadline.

By May 16 officials settled on three possible fixes:

- A sunshade, or parasol, that would be manually deployable through the Skylab's scientific airlock
- A twin-pole window shade-type device that would be deployed by booms
- An inflatable mattress-type shield

As a contingency, NASA was designing an improved thermal shield, which they dubbed a SEVA sail (SEVA is the acronym for "standup extravehicular activity"). The SEVA sail would be rigged with ropes from the hatch of the command service module (CSM).

The first manned visit to Skylab was rescheduled for May 25. On May 23 NASA made its choices: use of the parasol would be primary, the window-shade device secondary, and the SEVA sail would be the contingency. Meanwhile, the astronauts were rehearsing maneuvers to see if they could free the remaining solar array, which was believed to be stuck. They stored the most versatile of the cutting and prying tools in the CSM in preparation for the launch.

Astronauts repair crippled space station

Astronauts Pete Conrad, Paul Weitz, and Joe Kerwin rocketed away from Earth on May 25, the first manned launch to the orbiting space station. They saw for themselves that the shield and solar array were missing. The next day the astronauts made history as they docked with Skylab, entered it, and successfully deployed the parasol. Interior temperatures immediately decreased and stabilized at about 80°F.

More repairs were to follow. On June 7 Conrad and Kerwin released a restraining clamp and cut the embedded meteoroid shield joint, which had prevented the remaining solar array from opening. As it gradually and fully extended, Skylab's capacity to generate electrical power doubled to almost 7 kilowatts (kW). The solar arrays were originally designed to provide 10.5 kW, but NASA experts felt 7 kW was an acceptable compromise. In fact, temperature and power supplies remained satisfactory for the rest of this first manned trip and for the remainder of the program's missions. Skylab had been dramatically saved from the brink of disaster.

Impact

In light of the near-disaster, however, a Skylab Investigation Board was convened, chaired by Bruce T. Lundin, director of NASA Lewis Research Center. The board held consultations and inquiries at various locations, then reported to NASA administrator James C. Fletcher on July 13, 1973.

Investigators blame NASA management for the failure

The board uncovered what it regarded as a major design omission: the meteoroid shield was not designed to cope with the aerodynamic loads that the shield would be subjected to during the launch phase.

Ample attention had been devoted to getting the shield to deploy in orbit, but none to ensuring that it would stay in place during flight. The board concluded that communications had broken down between the different NASA departments—aerodynamics, structures, manufacturing, and assembly personnel and systems engineering approval. Management procedures had not identified these problems, the board complained.

For future complex endeavors, the board advised NASA to strengthen management processes by guaranteeing that:

- Design engineers have more firsthand experience in testing, operations, and failure analysis. This would enable them to better conceptualize how hardware behaves and fails.

- Project engineers oversee all technical activity.

- Chief engineers, who should not be occupied with administrative or managerial duties, ensure integration of all elements of the system.

Skylab eventually hosted 9 astronauts for 171 days

Despite the close call, Skylab achieved and surpassed all expectations. The project was visited by 9 astronauts in 3 different crews who stayed there for periods of 28, 59, and 84 days. During those manned missions, the space station orbited Earth some 3,900 times at an average altitude of 270 miles (434.5 kilometers). Its orbital path took it over 75 percent of Earth's surface before it finally left its orbit, as planned, and burned up in Earth's atmosphere over the Indian Ocean on July 11, 1979. Until the Skylab Program, NASA had little actual experience with prolonged weightlessness, and the long-duration flights of Skylab provided NASA with invaluable data for future space flights.

Where to Learn More

Belew, L. F., ed. *Skylab: Our First Space Station, 1977.* National Aeronautics and Space Administration, Scientific and Technical Information Office. Washington, DC: Government Printing Office, 1977.

Bulban, E. J. "Skylab Array Deployed in Successful EVA." *Aviation Week and Space Technology* (June 11, 1973): 26–28.

Fink, D. E. "Skylab Parasol: Six-month Job in 6 Days." *Aviation Week and Space Technology* (June 11, 1973): 53–56.

George C. Marshall Space Flight Center. Skylab Program Office.

"MSFC Skylab Mission Report—Saturn Workshop, 1974." *NASA Technical Memorandum, NASA TM X-64814.* Washington, DC: National Technical Information Service, 1974.

National Aeronautics and Space Administration. *NASA Investigation Board Report on the Initial Flight Anomalies of Skylab 1 on May 14, 1973.* Washington, DC: NASA, July 13, 1973.

"Saving a Station in Space." *Science News* (May 26, 1973): 336–337.

Schneider, W. C., and W. D. Green Jr. "Saving Skylab." *Technology Review* (January 1974): 42–53.

"Troubled Skylab: A Mission in Jeopardy." *Science News* (May 19, 1973): 320–321.

U.S. House. Committee on Science and Astronautics. Subcommittee on Manned Space Flight. *Skylab 1 Investigation Report: Hearing, August 1, 1973.* 93rd Cong., 1st sess., 1973.

U.S. Senate. Committee on Aeronautical and Space Sciences. *Skylab: Hearing, May 23, 1973.* 93rd Cong., 1st sess., 1973.

———. *Skylab, Part 2: Report of Skylab 1 Investigation Board: Hearing, July 30, 1973.* 93rd Cong., 1st sess., 1973.

Wilford, J. N. "Skylab, Short of Power, Overheating: Further Delay for Astronauts Feared." *New York Times* (May 16, 1973): 1.

Challenger explodes

Cape Canaveral, Florida
January 28, 1986

Just 73 seconds into its launch on January 28, 1986, the space shuttle *Challenger* exploded. This mission, the twenty-fifth in the shuttle program, was the first ever to include a teacher—Christa McAuliffe—who was to broadcast lessons from space. There to witness the heartbreaking disaster were thousands of National Aeronautics and Space Administration (NASA) personnel, relatives and friends of the crew, journalists and broadcasters, and spectators. The seven-person crew was inside a module that detached from the shuttle during the blow-up. Evidence indicates that the crew members survived the explosion, only to die—after a nine-mile free fall—when their craft plunged into the Atlantic Ocean.

When Challenger *explodes, seven astronauts die. Government investigations and shuttle redesign grounds the shuttle program for well over two years.*

Background

"The main rule about manned flight in space . . . is [that it is] dangerous and always potentially deadly," wrote Joseph J. Trento in his book *Prescription for Disaster.* Safety of the astronauts was an overriding concern at the beginning of NASA, which was created on July 29, 1958. President Dwight D. Eisenhower appointed T. Keith Glennan as NASA's first administrator, and Glennan brought on Robert R. Gilruth to coordinate Project Mercury, which was commissioned to get men into space.

Gilruth exemplified the agency's concern for safety. Trento noted that "Gilruth befriended the astronauts and suffered over the fact that their lives were in his hands. Gilruth would not go on with Mercury unless he was absolutely certain his astronauts could get off an exploding Atlas. He

During its minute in flight, *Challenger* exhibited no problems or complications.

had little faith in the Atlas rocket and dreaded the idea that one of 'his boys' might be blown to bits on his command." Trento reported that "with Mercury, a tradition started at NASA in manned flight. A tradition that precluded any budget savings in areas of mission safety. No part of Mercury, Gemini, and later Apollo got more attention than escape systems and astronaut safety. Only in 1972, with the approval of the Space Shuttle design, did NASA abandon that tradition."

U.S. is hysterical as Russians launch Sputniks

But NASA was not created simply to realize manned space exploration. It was borne out of Cold War hysteria. On October 4, 1957, the Soviet Union (USSR; now the former Soviet Union) used the world's first Intercontinental Ballistic Missile (ICBM) to launch *Sputnik I,* an orbiting spy satellite that passed over the United States. *Sputnik II* went into orbit November 3, 1957; *Sputnik III* was sent up May 15, 1958. If the Soviets could get satellites into earth orbit, then they could also arm their ICBMs with nuclear warheads and launch them onto the United States.

The creation of NASA as a civilian government agency was a deliberate effort on President Eisenhower's part to temper the hysteria and competition that were growing over the spy satellites and other Soviet space achievements. As Trento noted: "Eisenhower knew that . . . Congress would never allow such a Soviet feat [launching of the *Sputnik* satellites] to go unanswered. Eisenhower was adamant that the military not be given additional programs and power. And that meant that he was not going to put a manned space program in the Pentagon even though military research had yielded interesting information on what man might find in space. Eisenhower became convinced that only a civilian agency could keep interservice rivalry from hampering manned space research."

Air Force, wanting space to be top secret, stifles scientists

Eisenhower's fears about the military were not groundless. The Air Force was researching space activity since the late 1940s, and by the 1950s all three branches of the military had separate programs. After World War II, the United States rescued 180 of Germany's most brilliant rocket scientists, including Wernher von Braun. However, interservice competition had a stifling effect. Von Braun worked for an Army rocket team, and the Air Force prevented von Braun's group from building a rocket that could orbit a satellite. Instead, the Naval Research Laboratory was given the project. As Trento related: "The Soviets captured 6,000 scientists and mechanical engineers and launched the rockets von Braun and his U.S.-captured colleagues had conceived and designed. The Soviets went to work immediately launching captured V-2 rockets and refining the technology. But von Braun and his group got far less encouragement and support."

According to Trento, the military services resented a civilian agency exploring space, and "that resentment would haunt the agency for its entire history." Indeed, NASA had to depend on the military to get work done. They needed the Air Force for rockets and their bases for launches. They needed the Navy to rescue the space capsules in the sea. For such services, the military required concessions from NASA from the very beginning. By the time the space shuttle program was being developed, the Air Force insisted it could bump any civilian cargo at any time.

Apollo, NASA quality declines

Space exploration is vastly expensive, and the money NASA used—public money—was greatly affected by the winds of political change.

Funding for NASA depended on the commitment of the political party that held sway as well as whatever domestic and world situations that drained financial resources, such as the Vietnam war. In the 1970s and 1980s, pressure to balance the budget—and the subsequent cuts in funding—further eroded NASA's ability to monitor safety and control quality. "The space agency that was a remarkably effective and very special organization continued to deteriorate," Trento wrote. "By the end of 1980, NASA's capability to technically verify any contractor's work had all but vanished. By the end of 1980, the NASA that was once the toughest quality-control operation in or out of government was now depending on the military for many of its inspections. And by the end of 1980, the NASA veterans who fought the hardest for safety were leaving."

Then in 1982, a National Security Decision Directive dictated that NASA's highest priority was to make the shuttle program "fully operational and cost-effective in providing routine access to space." To help achieve cost effectiveness and be less dependent on the military, NASA began to "hire out" the shuttle's cargo bay to businesses to deliver their payloads into space. However, the military and certain politicians were opposed to such commercial use of the shuttle.

NASA hopes to make 1986 a breakthrough year

Subsequent National Security directives called for scheduling as many as 24 flights a year, of which the Air Force would reserve 6 for its own exclusive use. Abandoning rockets, which were used only once and then discarded, the Air Force decided to launch all U.S. satellites by shuttle. This put enormous pressure on NASA, which previously planned to use expendable rockets as a backup to the shuttle for satellite launching. By the end of 1985, NASA had a dismal record—no more than 9 missions flew in any given year. Greatly underestimating the turnaround time between scheduled launches to be 160 hours, NASA discovered that they needed 1,240 hours minimum. The $9 billion shuttle-based space transportation system was turning out to be neither cost efficient nor reliable.

With pressure mounting to meet an impossible schedule, NASA decided 1986 would be the shuttle program's breakthrough year. In January it announced an ambitious schedule of 15 missions using all 4 of its shuttles. The schedule had to be implemented immediately in order to realize its goal of more than one per month, but technical delays interfered immediately. After at least seven separate postponements, *Columbia* flew the year's first shuttle mission, 61-C, on January 12. Bad weather prolonged its stay in space, and by the time *Columbia* returned to Earth January 18, NASA's 1986 schedule was already in jeopardy.

Challenger picked to fly the year's second mission

Meanwhile *Challenger* had to be readied for the second January mission. *Challenger* had last flown on November 6, 1985. This latest mission featured the much-publicized Teacher-in-Space broadcasts as well as plans to launch a Tracking Data and Relay Satellite (TDRS). During the six-day mission, *Challenger* would also recover the Spartan-Halley comet research observatory from orbit. The Spartan's launch window requirement—it had to be orbited no later than January 31—was inflexible, as was the sequence of missions in line to follow. The tight scheduling led NASA to prepare contingency plans to skip the *Challenger* mission if it could not fly by the end of the month.

The Spartan project scientists preferred an afternoon launch, but NASA decided on mid-morning. NASA knew that if the shuttle suffered an "engine-out" during launch, the shuttle would use its planned emergency landing site at Casablanca, on the west coast of Africa. An afternoon launch off Florida meant that an emergency landing in Africa would occur at night. NASA wanted to avoid that scenario, because the Casablanca runway was not equipped with lights. (In any case, weather at Casablanca was reported to be poor.)

On January 15, 1986, NASA held its *Challenger* Flight Readiness Review. Teleconferencing with all the centers involved in the project, NASA reviewed all systems in detail, from the engineering of the spacecraft to the in-flight responsibilities of Johnson Space Center in Houston, Texas, and Marshall Space Flight Center in Huntsville, Alabama. The conference concluded with a "Go for launch."

Crew includes first-ever Teacher in Space

The seven-person crew chosen for the mission was commanded by Francis Scobee, who had piloted a 1984 shuttle mission. This would be the first time in space for his pilot, Michael Smith, and for the payload specialist in charge of the TDRS satellite, Gregory Jarvis. Mission specialists Ellison Onizuka, Ronald McNair, and Judith Resnick, who ran the satellites and experiments, were all experienced space travelers. The crew also included another rookie at space travel, the 37-year-old Christa McAuliffe from Concord High School in New Hampshire, the Teacher in Space.

Scheduled for January 22, the mission was first postponed to January 24, then to January 25. A forecast of bad weather for January 26 held up the mission until Monday, January 27, when a problem with a hatch bolt suddenly developed. By the time this problem was corrected, crosswinds had built up to a dangerous 30 knots. Although the crew was ready to launch and the shuttle had been fueled, liftoff had to be rescheduled for Tuesday, January 28.

Details of the Disaster

That night, Cape Canaveral temperatures dropped to well below freezing. NASA managers and contractors met for a late-night review. They were becoming increasingly concerned about the cold weather. No shuttle had ever been launched at temperatures below 53°F. Engineers from Morton Thiokol, a NASA contractor, warned that the O-rings on the shuttle's solid rocket boosters stiffen in the cold and lose their ability to seal properly. NASA managers pushed for a go or no-go, and Morton Thiokol managers, overruling their own engineers, signed a waiver stating that the solid rocket boosters were safe for launch at the colder temperatures.

Having decided to go ahead with the launch, NASA turned its attention to the ice on the shuttle and launchpad, which formed as temperatures dipped from 29° to 19°F. There were actually icicles up there! If they broke off during launch, they could damage the insulating tiles, which protect the shuttle as it reenters Earth's atmosphere. NASA rescheduled the launch from 9:38 A.M. to 10:38, and then to 11:38. Meanwhile, inspection teams surveyed the craft's condition and reported that the ice build-up had caused no apparent abnormalities. That was the report NASA managers were waiting to hear. Finally, at precisely 11:38:00.010 EST, *Challenger* lifted off from Complex 39B at Cape Canaveral, to begin its tenth flight into space.

Plume of flame visible at 59 seconds

Cheering and celebrating the majestic launch was a crowd of spectators, including Christa McAuliffe's family and even a group of her pupils. As *Challenger* rose into a clear, cold blue sky, no one on the ground or in the shuttle realized that a fire flamed out of the right-hand booster rocket, jetting down toward the giant fuel tank. The vehicle then rolled to align

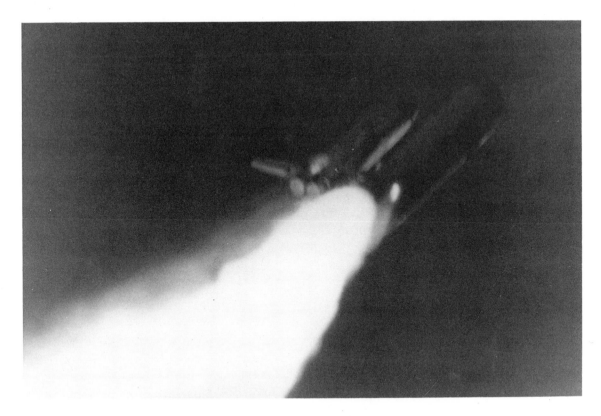

At 63.47 seconds after launch, the fire jetted down from the right-hand booster rocket and began to extend over the surface of the giant fuel tank.

itself on the proper flight path and throttled back its engines. The plume of flame became evident about 59 seconds into the launch.

By 64 seconds into launch, the fire burned a gaping hole in the casing of the booster. At 72 seconds, it loosened the strut that attached the booster to the external tank. "Uh oh," uttered pilot Michael Smith, the only evidence on the cockpit's flight recorder that anyone onboard suspected any trouble. One second later the loosened booster rocket slammed into the tip of *Challenger's* right wing. Then, at an altitude of 46,000 feet, the booster rocket crashed into the fuel tank and set off a massive explosion. The shuttle was traveling about twice the speed of sound.

At first the accident seemed a spectacular separation of the booster rockets. But when the fireball widened and debris began to scatter, even the spectators knew something was wrong. Everyone fell silent, stunned. The shuttle could no longer be seen.

The crew had no survivable abort options

NASA began rescue operations immediately, but the chance of finding survivors was very remote. Once the solid rocket boosters were ignited, the crew had no survivable abort options. If something went wrong at that critical moment of the launch, there was nothing anyone could do.

Challenger exploded 20 miles off the coast of Florida. The force of the explosion sent debris 20 miles above the earth. Burning fragments of the shuttle rained down on recovery operations for the next hour. Of all the accidents in the 25-year history of manned spaceflight, the *Challenger* disaster was by far the worst and marked the first time that American astronauts lost their lives during a mission. The disaster, viewed continuously on television, sent shock waves through the nation.

Impact

President Ronald Reagan eulogized the *Challenger* crew during the memorial ceremony at the Johnson Space Center in Houston, Texas. Then, on February 3, 1986, Reagan established a presidential commission to investigate the accident. He appointed William B. Rogers, the former secretary of state, to chair the commission.

Crew died as module crashed into ocean

The shuttle's crew module was recovered from the Atlantic ocean floor six weeks after the disaster, and the crew members were buried with full honors. Had the crew survived the initial explosion? Considerable speculation centered on this question, until NASA released its findings. Evidence indicates that the crew did indeed survive breakup and separation and had initiated emergency procedures. It is unknown if the entire crew remained conscious throughout the 2¾-minute free fall into the ocean, but at least two crew members activated emergency air packs.

Although the crew module has never been exhibited publicly, photographs of the cabin that were released showed nothing recognizable. Experts estimate that the module struck the surface of the ocean at a speed of nearly 2,000 miles per hour. The 16½ foot high cabin was compressed into a solid mass half the original size, which would certainly have killed anyone still alive in the module. The module's thick windows were shattered, but there was no evidence of fire.

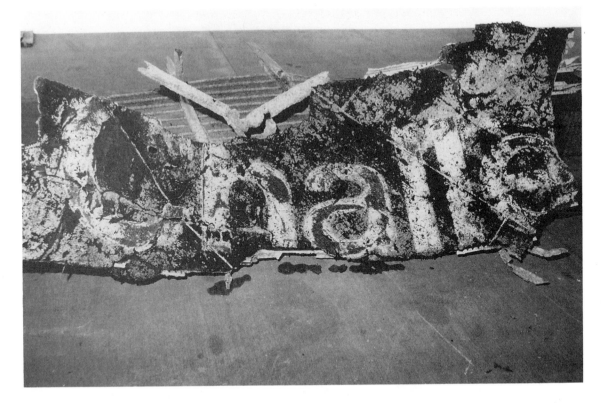

This segment of the *Challenger*'s right wing was recovered, along with other debris, by Navy divers about 12 nautical miles northeast of Cape Canaveral in 70 feet of water.

Rogers Commission investigates *Challenger*

The Rogers Commission assessed NASA and the *Challenger* incident by conducting a three-month investigation that involved more than 6,000 people. The commission recorded 15,000 pages of testimony during public and closed hearings, collected 170,000 pages of documents as well as hundreds of photographs, and sponsored independent technical studies. In its methodical review of the events, the commission also evaluated flight records, film evidence, and recovered debris.

Commission's findings indict O-rings

The Rogers Commission released its findings on June 6, 1986. It determined that the immediate physical cause of the *Challenger* disaster was "a failure in the joint between the two lower segments of the right Solid Rocket Motor," and specifically "the destruction of the seals that are intended to prevent hot gases from leaking through the joint during the propellant burn." In attributing the disaster to the destruction of these seals, the commission focused on the O-rings. The shuttle's large strap-on booster rockets were built in four sections; rubber O-rings sealed the sections together. Zinc chromate putty was supposed to keep the hot combustion gases inside from coming into contact with the rubber rings.

O-rings had history of partial failures

When it checked into the history and performance of this O-ring sealing system, the Rogers Commission was shocked to learn that the O-rings had failed regularly—even if only partially—on previous shuttle flights. Although concerned about the frailty of the seals, NASA and Morton Thiokol decided not to redesign the system. Because the seals had never failed completely, both had considered O-ring erosion an acceptable risk. But the frigid temperatures on *Challenger*'s January 28 flight made the O-rings less resilient, or less flexible, than normal, and they did not completely seal the joint. Photographs reveal that even before the shuttle had cleared the launch tower, hot gas was already "blowing by" the O-rings.

NASA and Morton Thiokol blamed for flawed decision

The Rogers Commission's 256-page report concluded that "the decision to launch the *Challenger* was flawed." The commission blamed the management structure of both NASA and Morton Thiokol for not allowing critical information to reach the right people. This assessment was seconded by the House Committee on Science and Technology, which spent two months conducting its own hearings. The congressional committee determined that the technical problem had actually been recognized early enough to prevent the disaster, but "meeting flight schedules and cutting cost were given a higher priority than flight safety."

How the Crash Influenced History

The public condemnation of NASA had grave consequences. The nation's confidence in NASA was shaken, and its astronauts were extremely disturbed. They had never been consulted or even informed about the dangers that the current sealing system exposed them to. Allowing astronauts and engineers a greater role in approving launches was one of the nine recommendations the Rogers Commission made to

NASA. The commission's other recommendations included a complete redesign of the solid rocket boosters joints, a review of the astronaut escape systems towards achieving greater safety margins, regulation of the rate of shuttle flights to maximize safety, and sweeping reform of the shuttle program's management structure.

Long hiatus before shuttle program flies again

In the wake of the commission's findings, several key people left NASA, including a number of experienced astronauts. They resigned, disillusioned with NASA and—given the long redesign process now pending—frustrated that there would be even fewer chances to fly. The shuttle program was earthbound for more than two years. The twenty-sixth mission did not leave the launchpad until September 29, 1988, when—to everyone's relief—*Discovery* flew without incident. NASA also built a shuttle to replace *Challenger;* this craft, *Endeavour,* first flew on May 7, 1992.

NASA must rebuild its credibility

The shuttle no longer needs to be the sole deployer of unmanned satellites. After the *Challenger* explosion, NASA has been launching small American and foreign satellites with expendable rocket vehicles. But public perception of NASA has eroded. NASA's funding problems, particularly with its new space station plans, indicate the uphill struggle the agency faces to regain its status.

Where to Learn More

Bell, Trudy E., and Karl Esch. "The Fatal Flaw in Flight 51-L." *IEEE Spectrum* (February 1987): 36–51.

Boffey, Philip M. "NASA Had Warning of a Disaster Risk Posed by Booster." *New York Times* (February 9, 1986): 1.

Broad, William J. "Thousands Watch a Rain of Debris." *New York Times* (January 29, 1986): 1.

DeWaard, E. John, and Nancy DeWaard. *History of NASA: America's Voyage to the Stars.* Rev. ed. New York: Exeter Books, 1987.

"A Fatal 'Error of Judgment.'" *Newsweek* (March 3, 1986): 14–19.

Lewis, Richard S. *"Challenger": The Final Voyage.* Columbia, 1988.

McConnell, Malcolm. *"Challenger": A Major Malfunction.* New York: Doubleday, 1987.

"NASA Identifies Failure Scenarios." *Aviation Week and Space Technology* (March 17, 1986): 25–26.

Presidential Commission. *Report of the Presidential Commission on the Space Shuttle "Challenger" Accident.* 5 vols. Washington, DC: Government Printing Office, June 6, 1986.

Sanger, David E. "Shuttle Changing in Extensive Ways to Foster Safety." *New York Times* (December 28, 1986): 1.

Smith, Melvyn. *Space Shuttle.* Newbury Park, CA: Haynes, 1989, pp. 264–285.

Smith, R. Jeffrey. "Inquiry Faults Shuttle Management." *Science* (June 20, 1986): 1,488–1,489.

Trento, Joseph J. *Prescription for Disaster: From the Glory of Apollo to the Betrayal of the Shuttle.* New York: Crown, 1987.

Bibliography

Barclay, Stephen. *The Search for Air Safety: An International Documentary Report on the Investigation of Commercial Aviation Accidents.* New York: William Morrow, 1970.

Bignell, Victor, Geoff Peters, and Christopher Pym. *Catastrophic Failures.* Bristol, PA: Open University Press, 1977.

Bishop, R. E. D. *Vibration.* Winchester, MA: University Press, 1965.

Blockley, D. E. *The Nature of Structural Design and Safety.* Ellis Horwood, 1980.

Canning, John, ed. *Great Disasters.* Stamford, CT: Longmeadow Press, 1976.

Collins, J. A. *Failure of Materials in Mechanical Design: Analysis, Prediction, Prevention.* New York: Wiley, 1981.

Cornell, James. *The Great International Disaster Book.* New York: Scribner's, 1976.

Davis, Lee. *Man-Made Catastrophes: From the Burning of Rome to the Lockerbie Crash.* New York: Facts on File, 1993.

Ebert, Charles H. V. *Disasters: Violence of Nature, Threats by Man.* Dubuque, IA: Kendall/Hunt, 1988.

Eddy, Paul, Elaine Potter, and Bruce Page. *Destination Disaster: From the Tri-Motor to the DC-10; The Risk of Flying.* Quadrangle, 1976.

Editors of Encyclopaedia Britannica. *Catastrophe! When Man Loses Control.* New York: Bantam/Britannica Books, 1979.

Feld, Jacob. *Construction Failures.* New York: Wiley, 1968.

——. *Lessons from Failures of Concrete Structures.* Detroit, MI: American Concrete Institute, 1964.

Florman, Samuel C. *Blaming Technology: The Irrational Search for Scapegoats.* New York: St. Martin's Press, 1981.

Ford, Daniel. *O-Rings and Nuclear Plant Safety: A Technological Evaluation.* Washington, DC: Public Citizen, Critical Mass Energy Project, 1986.

Frank, Beryl. *Great Disasters of the World.* Austin, TX: Galahad Books, 1981.

Godrey, Edward. *Engineering Failures and Their Lessons.* Privately printed, 1984.

Godson, John. *Unsafe at Any Height.* New York: Simon & Schuster, 1979.

Gordon, J. E. *The New Science of Strong Materials; or, Why You Don't Fall through the Floor.* 2d ed. New York: Penguin Books, 1976.

———. *Structures; or, Why Things Don't Fall Down.* New York: Da Capo Press, 1981.

Great Britain Navy Department Advisory Committee on Structural Steels. *Brittle Fracture in Steel Structures.* Markham, Ontario: Butterworths Canada, 1970.

Guide to Investigations of Structural Failures. Washington, DC: Federal Highway Administration, 1980.

Hammond, Rolt. *Engineering Structural Failures: The Causes and Results of Failure in Modern Structures of Various Types.* Odhams Press, 1956.

Hertzbert, R. W. *Deformation and Fracture Mechanics of Engineering Materials.* New York: Wiley, 1976.

Janney, Jack R. *Guide to Investigation of Structural Failures.* New York: American Society of Civil Engineers, 1979.

Keylin, Arleen, and Gene Brown. *Disasters: From the Pages of the New York Times.* New York: Ayer, 1976.

Kletz, Trevor A. *What Went Wrong? Case Histories of Process Plant Disasters.* Houston, TX: Gulf, 1985.

Launay, A. J. *Historic Air Disasters.* London: Ian Allen, 1967.

LePatner, Barry B., and Sidney M. Johnson. *Structural and Foundation Failures: A Casebook for Architects, Engineers, and Lawyers.* New York: McGraw-Hill, 1982.

Lewis, Elmer Eugene. *Nuclear Power Reactor Safety.* New York: Wiley, 1977.

Mair, George. *Bridge Down: A True Story.* Briarcliff Manor, NY: Stein & Day, 1983.

McClement, Fred. *It Doesn't Matter Where You Sit.* New York: Holt, 1969.

———. *Jet Roulette: Flying Is a Game of Chance.* New York: Doubleday, 1978.

McKaig, Thomas K. *Building Failures: Case Studies in Construction and Design.* New York: McGraw-Hill, 1963.

Nash, Jay Robert. *Darkest Hours*. Chicago: Nelson-Hall, 1976.

Oberg, James E. *Uncovering Soviet Disasters*. New York: Random House, 1988.

Osgood, Carl C. *Fatigue Design*. 2d ed. Elmsford, NY: Pergamon Press, 1982.

Perrow, Charles. *Normal Accidents: Living with High-Risk Technologies*. New York: Basic Books, 1984.

Petroski, Henry. *Design Paradigms*. New York: Cambridge University Press, 1994.

———.*To Engineer Is Human*. New York: St. Martin's Press, 1985.

Ross, Steven S. *Construction Disasters: Design Failures, Causes, and Prevention*. New York: McGraw-Hill, 1984.

Salvadori, Mario. *Why Buildings Stand Up: The Strength of Architecture*. New York: McGraw-Hill, 1982.

Salvadori, Mario, and Matthys Levy. *Why Buildings Fall Down: How Structures Fail*. New York: Norton, 1992.

Serling, Robert J. *Loud and Clear: The Full Answer to Aviation's Vital Question, Are the Jets Really Safe?* New York: Doubleday, 1969.

Stewart, Oliver. *Danger in the Air*. New York: Philosophical Library, 1958.

Turner, Barry A. *Man-Made Disasters*. New York: Crane, Russak, 1978.

U.S. House of Representatives. Committee on Science and Technology. *Structural Failures: Hearings before the Subcommittee on Investigations and Oversight*. Washington, DC: Government Printing Office, 1984.

———.*Structural Failures in Public Facilities*. Washington, DC: Government Printing Office, 1984.

Webb, Richard E. *The Accident Hazards of Nuclear Power Plants*. Amherst: University of Massachusetts Press, 1976.

Whyte, R. R., ed. *Engineering Progress through Trouble*. Institution of Mechanical Engineers, 1975.

Index

Bold denotes entries.